Technology Adoption in the Caribbean Tourism Industry

Andrew Spencer

Technology Adoption in the Caribbean Tourism Industry

Analyzing Service Delivery in the Digital Age

palgrave
macmillan

Andrew Spencer
Tourism Product Development Company
Kingston, Jamaica

ISBN 978-3-030-61583-3 ISBN 978-3-030-61584-0 (eBook)
https://doi.org/10.1007/978-3-030-61584-0

This Palgrave Macmillan imprint is published by the registered company Springer Nature Switzerland AG
The registered company address is: Gewerbestrasse 11, 6330 Cham, Switzerland

Contents

List of Figures

List of Tables

CHAPTER 1

Introduction

According to the World Travel and Tourism Council (WTTC), the tourism industry in the Caribbean contributes to 13.9% of the region's gross domestic product (GDP).[1] Consequently, the state of the tourism industry is exceptionally important to the Caribbean. However, the sector is remarkably vulnerable to both internal and external shocks; including natural disasters, global pandemics, economic crisis among other factors. In light of the sector vulnerability there is a need to mitigate the impact of these vulnerabilities, especially in light of the most recent global pandemic—Covid-19. Which has significantly, reduced the movement of people across the world, which has tremendous economic impact on the Caribbean key foreign exchange sources. The use and widespread integration of Information communication Technologies has the ability to mitigate the challenges experienced by the industry and improve the provision of service.

As technology develops and firms increase their adoption of Information and Communication Technologies (ICT), there comes a shift in focus from what one may call "the Big Picture" to more bespoke solutions that are appropriate for diverse, individualized scenarios. This is not to say that more orthodox or mainstream management practices

[1] The World Tourism and Travel Council. (2019). Economic Impact Reports. Retrieved from https://wttc.org/Research/Economic-Impact.

A. Spencer, *Technology Adoption in the Caribbean Tourism Industry*,
https://doi.org/10.1007/978-3-030-61584-0_1

have been discarded. While these still prove effective, especially in smaller owner-managed travel firms, it does illustrate an expansion of perspectives.

The use of computers and other technologies in the travel industry is not a new idea. Computers have been in use by travel agencies since the 1950s with the creation of the first reservation systems—TIS and Gulliver during the 1980s,[2] and major global distribution systems (GDSs) like SABRE and Amadeus in the 1990s. Over the years these systems have evolved into more advanced tourist information systems.

We see this trend continuing today, resulting in a myriad of travel reservation and monitoring systems across the internet; each of them able to access, and benefit from, the others' information databases and architectures. This has resulted in the most leveled playing field the industry has witnessed to date, with numerous options in products and services available to both producers and consumers.[3]

Emerging characteristics of this trend include the development of new value chains and systems. This new infrastructure allows industry players to take increased advantage of the opportunities presented by applying the latest technologies.[4] However, in order for a travel agency to take advantage of these benefits, certain preconditions must be met.

Moital et al. (2009) make it clear that the adoption of this new paradigm requires familiarity with, or at least a basic understanding of, the tools inherent in this new paradigm. Specifically, computers and the internet.

A travel firm's staff requires a certain minimal level of fluency in computer literacy. They may not need to be able to code programs, but they do need to understand how to interact with computer hardware and software. This is a fundamental prerequisite to being able to access and navigate the internet.

The staff also require, at the very least, a basic level of understanding of the internet itself. This is necessary for them to be able to effectively engage in activities such as internet sales, also referred to as e-commerce, and internet marketing. These may also be referred to as ***SEM*** (Search Engine Marketing) and ***SEO*** (Search Engine Optimization). Internet

[2] Werthner (1995, 1996), Werthner and Klein (1999).

[3] Poon (2001).

[4] Buhalis (2002).

Marketing is a wider term, encompassing both SEM and SEO, but in common parlance, they tend to be used interchangeably.

This stepwise progression allows us to appreciate the dependencies between each step. This also establishes the notion that engagement at each step assumes previous engagement (and experience) in preceding steps. We may refer to this as the "innovation interdependence perspective." From this perspective, we can project a full stack of progressive steps as follows:

Since 2000 it's been argued that different regions in the world are at different levels when it comes to being able to integrate and utilize the internet; this is referred to as "internet readiness." While America and Canada were strong early adopters, Europe, in the nineties and early 2000s, had been a year and a half behind North America when it came to levels of internet adoption.[5] However, while Europe has essentially caught up to North America, countries in the developing world, including the Caribbean, are lagging behind, and there are unfortunate consequences for countries and businesses that are slow to integrate ICTs.

Most travel agencies in the Caribbean are focused on the outbound traveler. As such, they may not experience issues like disintermediation—the consolidation and reduction of intermediaries—at the same pace as more developed countries. This has resulted in a lack of strategic planning and proactive measures on the part of local travel agencies.

The reason for this was partially the perception that most Caribbean people did not place much trust in the concept of making purchases on the internet, reflected in a study by Lin.[6] This led to certain considerations. Firstly, most travel firms in the Caribbean are focused on the outward-bound market. Secondly, there is a gap with regard to access to technology (what may be termed as the "digital divide") between tourists and Caribbean travel destinations. These considerations raise the question of whether the inbound travel market is a gap that the local travel firms can fill if they can bridge the digital divide.

There is, however, another disincentive for local firms in catering to the inbound market. Inbound travel does not provide a commission comparable to what they may make in outbound travel. This is, unfortunately, a shortsighted view as it fails to take into consideration opportunities

[5] Law and Leung (2000).

[6] Lin et al. (2009).

presented through value-adding and dynamic packaging, i.e., creating travel solutions on a more individualized basis.

More recently, the digital divide has been shrinking, but according to Minghetti and Buhalis,[7] there are still significant gaps between Caribbean destinations, developed nations and the tourists thereof. This will lead to varying levels of "digital exclusion," where certain companies lose business opportunities because of some technological inadequacy.

Minghetti and Buhalis have conducted research on the digital divide in the context of the tourism industry. Their work highlights challenges in areas such as marketing and communications between tourism-generating countries and tourist destinations. Tourists and enterprises from these developed, tourism-generating regions interact on electronic platforms. These platforms reduce the need for brick-and-mortar locations where enterprises and clients have to interact face-to-face. According to Minghetti and Buhalis,

> The study of the digital divide is critical for less technologically developed regions that need to expand their ICT usage to be able to promote their offerings, interact with consumers, and reduce their dependence on intermediaries.[8]

One important consideration from the study is the fact that tourists who are not tech-savvy, and destinations that are behind current trends in ICT, still rely on the older way of doing things and, naturally, utilize travel agencies with physical locations. This scenario applies to the Caribbean.

There appears to still be a fairly high dependence on physical intermediaries, i.e., travel agencies with physical locations that clients can visit for face-to-face engagement. Specifically, The Jamaica Tourist Board, along with other enterprises in the tourism and hospitality industry, ensure that relationships are maintained with intermediaries such as The American Society of Travel Agents and The Association of British Travel Agents.

Due to this state of affairs, three issues become clear.

Tech-savvy tourists from developed countries who prefer to conduct their travel arrangements online may not be captured by more traditional marketing efforts in the Caribbean.

[7] Minghetti and Buhalis (2010).

[8] Minghetti and Buhalis (2010, p. 278).

The Caribbean's image as a tourist destination may be negatively affected if they are unable to interact with the islands' tourism and travel options via the internet. According to Govers, "covertly induced and autonomous agents, in particular, have a dramatic influence" over the image of a travel destination in the minds of consumers.[9] These agents include television, magazines, and the internet itself. And in the case of Jamaica, where the Jamaica Tourist Board is focusing most of its efforts on television advertisements,[10] online promotion and engagement are paid little attention; this is the reality for many other Caribbean islands as well. As a result, the television advert may intrigue potential tourists, but be disappointed when attempting to gain more information or seek booking options online. Such circumstances wear away at a destination's image.

Travel agencies within the Caribbean have significantly lessened over the past decade. Regardless, they do play an active role in driving inbound travel and tourism to the islands. They are typically focused on the outbound market due to the fact that they do not receive commission on inbound travel sales. The key to the issues previously highlighted the online presence of Caribbean islands and their travel firms. Currently, travel agencies receive little attention from local statutory bodies. The potential opportunities provide an excellent argument for greater governmental collaboration and support. The challenge is to motivate existing travel firms to adopt ICT beyond simply an exchange of emails, and to establish a stronger and more vibrant online presence. Right now, only about 5% of local travel companies have active websites. So, the question is, how do we motivate these companies to more readily adopt ICT and make it a pivotal component of their company strategy?

Challenges in the Industry

This work was motivated by the fact that the Caribbean travel industry is facing significant challenges. While the disintermediation that advanced economies already experienced is only recently taking place in the Caribbean travel industry, we have already seen a significant reduction in the number of firms. For instance, according to the Jamaican Association of Travel Agents (JATA), between 1999 and 2009 the number of

[9] Govers et al. (2007, p. 19).

[10] Williams and Spencer (2010).

travel agencies declined from 105 to 43; while in Trinidad and Tobago the number of agencies moved from 11 to 5.

Upon further assessment, many firms in Jamaica, The Bahamas, and Trinidad and Tobago were not implementing changes to similar operational procedures as those that were taking place in the global marketplace. The operational component that saw the least change was the use of technology.

After the adoption of Global Distribution Systems (primarily Sabre and Amadeus) in the 1990s, adoption of new technologies stalled. This was considerably more the case with technologies that directly impacted sales and marketing, despite the growing popularity of the internet. This observation leads one to wonder why domestic travel agencies were so slow to adapt despite the obvious need for evolution in their business models.

It would be beneficial to understand why firms with similar characteristics, and in similar contexts, have varying levels of ICT adoption. This research will look at a number of possible factors, with the aim of determining the prime factor(s), as well as investigating pertinent issues and challenges in the global environment.

Studies that focus on technology adoption are usually placed in one of two phases, pre-internet or post-internet. The most renowned *pre-internet phase* studies are based on earlier works by Rogers (1962) and Davis (1989). These two schools of thought have fundamental differences in understanding what drives the adoption of technology.

Roger's perspective, termed the "diffusion of innovations," focuses on innovation, communication, and the role of society. Davis, on the other hand, sees adoption from the perspective of the user and their assessment of this new technology. Is it easy to use? Does it do what I need it to? This school of thought is referred to as the "technology acceptance model."

When it comes to *post-internet phase* studies, the general consensus is that the internet is the most widespread, most pervasive technology ever devised. Even more so than those technological developments that led to improvements in farming production, such as those studied by Rogers (1962).

It has been observed over the last 30 years that industrial economies have evolved into information economies. According to Parker (1988), it is information, not land nor capital, that will drive the creation of wealth and prosperity for the foreseeable future. Technology has permanently changed the way that the world does business. Drucker (1990) takes it a step further by arguing that, due to this shift, knowledge now has the greatest impact as a means of production.

Naturally, those businesses that are quick to adopt technologies that facilitate the transfer of knowledge (ICTs) gain a competitive advantage.[11] Unfortunately, providing a simple definition of ICT is not an easy thing to do.

Buhalis (2002) shows that ICTs include hardware, software, groupware (software that allows multiple remote individuals to collaborate on a common project simultaneously), and NetWare (hardware and software that facilitates communication between computers, including other devices, on a digital network). But other major components of ICT are the capacities and capabilities of the users of ICTs to develop, program, and maintain these technologies. The best technologies in the world are only as powerful as the capabilities of those who are using them. Given how broad the topic of ICT is, it's important to point out that the main focus of this book is the single most pervasive technology of them all, the Internet.

While the internet was originally conceived and had its tentative start in 1969, it was not until 1991, more than twenty years later, that it became publicly accessible and grew into the internet we know today.

The internet became one of the most ubiquitous technologies ever, spreading across regions faster than any technology before it. Its impact and potential are so profound that organizations and businesses worldwide had to reassess their policies, procedures, and general ways of doing things.[12]

Other *post-internet phase* theorists argue that, apart from affecting various aspects of business, the internet has led to the modification and restructuring of entire economic sectors.[13] The internet demonstrated this amazing ability to penetrate and transform these multiple economic sectors and industries, both as an external force and as an internal driver.

[11] Porter (2001).

[12] Klein (1996), Grieger (2003), Amit and Zott (2001).

[13] Kalakoa and Whinston (1996), Gatty (1998), Ghosh (1998), Timmers (1998), Wirtz (2001).

The Research Context

In the Caribbean, travel and tourism is one of the leading sources of economic activity, as such the industry is one of the leading industry providers of jobs in Jamaica, and the Bahamas whether directly or indirectly. While in Trinidad it contributes significantly in regards to employment, with room for tremendous improvement. That being the case, it is vital that we gain an understanding of the local economic environment in which travel and tourism firms operate.

The Bahamas and Trinidad and Tobago can be classified as high-income countries while Jamaica is a middle-income country. Both Jamaica and The Bahamas are oil-importing countries. The typology of these Caribbean countries under examination is slightly nuanced with The Bahamas being a predominantly customer service export or tourism industry; Jamaica on the other hand can be classified as a mixed goods export/customer service (or tourism) industry while Trinidad and Tobago mainly exports Goods particularly, oil but the tourism industry is one that thrives in the country.[14] The Caribbean has made tremendous progress in the post-independence period, with numerous countries like the aforementioned earning middle income and high-income status based on growth per capita income. Nonetheless, the region is riddled with several developmental challenges which negatively impact growth and productivity. In light of these challenges, the region has been tasked to rethink and restructure their approach to development. Therefore, structural changes in terms of identifying and investing in priority economic industries especially when faced with crisis have been the balancing act carried out by Caribbean states. For instance, Jamaica's focus on the tourism industry occurred due to the downturn of the bauxite and aluminum industry in the 1970s despite it being the heart of the Jamaican economy decades prior.

The social context of the Caribbean influences the perception of new technologies. A major vehicle for this influence is the formal education system. Before independence, most schooling was provided by churches; today, the majority of schools are provided by the government.[15]

[14] Economic Commission for Latin America and the Caribbean (ECLAC). 2012. Development paths in the Caribbean. Retrieved from https://www.cepal.org/en/publications/38253-development-paths-caribbean.

[15] Whyte (1983).

Driven by a mandate of universal literacy, tuition became free and school attendance in the Bahamas, Jamaica and Trinidad and Tobago was made compulsory for all children at various educational levels.

As for higher education, courses are offered by numerous tertiary institutions, namely The University of the West Indies which has campuses located in The Bahamas, Jamaica, and Trinidad and Tobago.

Interestingly, even some American universities have recently begun offering courses locally in Jamaica.

Courses on tourism have been available at the University of the West Indies since the 1970s. Initially, students would enter their first year at one of UWI's three campuses (Mona, St. Augustine and Cave Hill) and complete their final two years of study in The Bahamas. Since 2006, however, the Mona campus initiated a full three-year course, allowing Jamaican students who wish to study tourism to do so in their own country.

Each of the aforementioned tertiary institutions has contributed, to some degree, to tourism and hospitality training of the country's workforce. In Jamaica, this effort saw significant growth when the College of Arts, Science and Technology became the University of Technology (UT). Unlike the courses offered on other campuses, those offered by UT have a more scientific focus, with the additional aim of imparting technical skills to its students. It appears, however, that the "trickle-down" of these skills to the rest of Jamaican society has been slow, and this is having a direct negative impact on business in the island; the same can be said for other Caribbean states.

Travel agencies tend to use technology that has been adopted by the public at large. This makes the technological context of the country important when it comes to understanding why travel agencies operate the way they do. Internet penetration within The Bahamas, Jamaica, and Trinidad and Tobago are 89,[16] 95, and 160%[17] respectively.[18]

As mobile and cell phones increased in popularity, the demand for landlines dropped precipitously. The number of landlines supplied dropped

[16] https://www.helgilibrary.com/indicators/mobile-phone-penetration-as-of-population/bahamas/#:~:text=Mobile%20phone%20penetration%20as%20a,than%20in%20the%20previous%20year.

[17] https://oxfordbusinessgroup.com/overview/three-company-liberalisation-has-brought-host-benefits-consumers-along-stronger-0.

[18] https://www.internetworldstats.com/carib.htm.

from over half a million to approximately three hundred thousand as of 2006 in Jamaica. The introduction of internet connectivity to the Caribbean, connecting States like the Bahamas, Jamaica, and Trinidad and Tobago to the rest of the world opened up a whole new set of possibilities: e-commerce.

As developing nations, the Bahamas, Jamaica, and Trinidad and Tobago has a lot to gain by participating in the global economy. America is our largest trading partner, and we are competing with the world. It is vital that we implement the infrastructure that will allow us to engage with those consumers on the same level as other nations around the world. Given the limited size and scope of these States, including its human capital, we have to develop quickly. Otherwise, we risk eventually facing record deficits. While tourism, agriculture, mining or energy currently provide a substantial portion of the revenue for these States, all of those local industries, by US standards, suffer from deficient technology.

The Bahamas, Jamaica, and Trinidad and Tobago have never been a major tourist-*generating* country, and most of the travel associated with the islands is inbound. Specifically, tourism began in Jamaica in the 1890s when the United Fruit Company had been experiencing a spate of excess capacity in their ships. Seeing an opportunity, they began offering cruises to Jamaica. Tourist hotels on the island soon followed, but it was not until after the Second World War that tourism in Jamaica began to flourish. The number of hotels tripled between 1945 and 1970, as investments at that time were benefiting from accelerated depreciation allowances.

Being geographical neighbors, most of the Bahamas, Jamaica, and Trinidad and Tobago's inbound travelers are from the United States. Likewise, most of their outbound travel is to the United States. This is reflected in the sizable diaspora in that country. As a tourist destination, though, these States receive far more tourists than it generates.

While it is commonly understood that Bahamians, Jamaicans, Trinidadians and Tobagonians travel to visit friends and family overseas, data on them traveling to other countries for vacation is practically nonexistent. This is probably due to the limited economic means of the average household. Despite the lack of data for outbound vacationers, the number of Caribbean citizens living in other countries would indicate that there is a demand for outbound travel for those who wish to visit family and friends overseas.

For instance, the Jamaica diaspora is huge! According to some estimates, there are as many people of Jamaican descent living outside of the island as within it. The largest concentrations of the diaspora are located in three countries:

- The United States
- Canada
- The United Kingdom.

As of 2008, there are roughly 637,000 second-generation Jamaican immigrants (those born in foreign countries) in the United States; 123,500 in Canada; and 150,000 in the United Kingdom, according to each country's statistical bureau.

Outbound travel in Jamaica and other Caribbean States is usually handled by travel agencies with physical locations; agencies that have been facing significant challenges over the past two decades. In 2000, American Airlines, British Airways, and the regional-carrier Air Jamaica cut commissions for regional travel agencies from 9 to 6%. According to the Jamaica Gleaner, American Airlines implemented the cuts first, then travel agencies responded by boycotting the sale of American Airlines tickets.

This action proved futile, however, as American Airlines (which accounted for 70% of the air traffic to the region) opted to offer direct bookings. The agencies' position was further weakened when, two months later, British Airways and Air Jamaica followed suit. In 2009, further cuts followed, taking commissions from 6 to 3%. This, and other changes in airline reservation practices, has led to the closure of numerous travel agencies and a proliferation in unregistered agencies. In Jamaica the Minister of Tourism stated, local unregistered agencies have no insurance nor protection for their clients. Some have even faced the consequences for selling travel packages that did not materialize.[19]

To provide a deterrent against the practice, the government passed legislation to increase the fine for operating an unregistered travel agency to a maximum of JMD$1,000,000.

The strategy being used up to this point, offering a limited set of services to outbound travelers, is failing. This research posits that, in addition to a reexamination of the market, current business practices and processes must also be reexamined in order to generate a more adaptable business model, especially with regard to technology.

[19] Jamaica Gleaner (2011).

Methodology

The aim of this research is to determine what drives (or hinders) the adoption of internet technologies for sales and marketing purposes in small, owner-managed travel firms in the Caribbean. To that end, we have the following objectives:

1. To examine those factors and drivers that determine the different levels of technology adoption in travel agencies.
2. To explore the relationship between leadership and technology adoption in owner-managed small firms.
3. To look at the influence of internal factors such as strategy and resources on technology adoption in owner-managed small firms.
4. To investigate issues such as the digital divide and culture to see how they affect technology adoption in owner-managed small firms.
5. To determine those characteristics of leaders and firms that lead to the various levels of technology adoption develop a useful model of technology adoption for owner-managed small travel firms.

The debate so far has existed mainly between *positivism* and *interpretivism*, and each school of thought has greatly influenced how research has been conducted. However, to achieve the aims of this work, *Critical Social Science* provides greater grounding.

Embedded within Critical Social Science is the idea that, by providing critique, objective truths may be uncovered. Truths that can help people and encourage them to take action.[20] Critical Social Science aims to supply individuals and societies with the tools to enact positive change. More importantly, it allows for the uncovering of underlying structures and systems. This "uncovering" provides greater context for understanding societies and economies, allowing people to both see and alter deeper structures. Fundamentally, the aim of this research is to dispel myths and provide actionable information that will empower people to change society for the better.

To do this, we must explore the Caribbean context in order to determine the conditions that local travel agencies are facing. To adopt a "one-size-fits-all" approach, one would have to assume that the drivers of

[20] Neuman (2006).

technology acceptance and the diffusion of innovation are universal. This research supports the idea that firms should never become complacent and should explore all possibilities while keeping all observable threats and opportunities under consideration.

This research also aims to explore the opinions and feelings of stakeholders in order to gain further insight. It proposes to explore the issues through the qualitative inquiry of a small sample in order to extract more universal conclusions. This research is rooted in *Critical Science Theory*, but more specifically, it subscribes to the concept of *Bounded Autonomy*, which argues that while there is some degree of subjectivity in human actions, those actions are bound within identifiable limits. In other words, while there are limits to what can be done, people are free to make whatever combination and/or permutations of choices they wish within those limits.

This approach blends determinism (*Positivism*) and voluntarism (*Interpretivism*) to show how social/economic structures and human agency interact. Previous research on technology adoption mostly utilized determinism/positivism.[21] The blended approach in this work assumes that people will only make choices based on what they think is possible within identifiable limits. These limits may be cultural or material. Cultural constraints refer to values, beliefs, and norms, while material constraints may be a matter of limited resources. With this in mind, the research attempts to explore the reality of Jamaican travel agencies and how they relate to other realities.

The decision to focus this study on travel firms was as a result of:

1. the significant decline in the number of firms and
2. the lack of research into the Caribbean travel industry.

The lessons learned here may help to understand other similar industries and Caribbean territories. The primary commonalities being small, owner-managed firms in relationship-oriented, developing economies. This research is being conducted on a qualitative basis, with an inductive approach; it is more concerned with the meanings, definitions and concepts at play than the statistics and measures. It also means that we will

[21] Davis (1989), Bagozzi (2007), Fuchs et al. (2010).

be using a small sample size to derive truth that is more generally applicable. Furthermore, given that the core data of our research is more of a textual nature than a numerical one, we will be using the Template Analysis techniques. Template Analysis derives common themes and narratives from the text data collected. This can be text produced or used in the context of the evaluation regardless of the evaluation activity, i.e., data that was not generated by the evaluation.[22]

Neuman (2006) argues that qualitative researchers should develop rudimentary working ideas and continually refine them throughout the data collection and analysis process, instead of taking abstract ideas and turning them into full theoretical definitions when studies are still in their early stages. As data is collected and analyzed new concepts are developed, formulates for major constructs are defined and the relationships among them are considered, further developing the general theory. The initiation and operation of this process comes before the conceptualization of the relevant theories, but conceptualization occurs organically during the execution of the research.

The execution of the research also involves describing how working ideas are developed during the observation and collection of data. This makes it more of an "after the fact" description than a pre-planned technique. This also illustrates how data and preconceived notions about what that data means can become constructs.

Reliability and validity are typical notions within positivism. While reliability is concerned with the extent to which results are consistent over time, validity is concerned with the extent to which a study actually measures what was intended.[23] While these two issues may never be fully resolved in this type of (qualitative) research, it is incumbent upon us to demonstrate that the research is still credible and trustworthy. In fact, according to Patton (2002), validity *determines* reliability in qualitative research.

The research design, while predominantly qualitative, also incorporated quantitative data that was analyzed in tandem with the qualitative data. This made the analysis more robust and provided a foundation for deeper qualitative inquiry. The interviews were semi-structured, and the questions were guided by both the objectives and a rudimentary conceptual

[22] Crabtree and Miller (1999).

[23] Golafshani (2003).

framework. While this did provide a deductive element to the study, the dominant inductive approach was validated through a detailed template analysis which identified common themes across cases. The results were universal and transferable conclusions that were applicable across contexts that shared the aforementioned commonalities:

- small, owner-managed travel firms
- located in similar developing countries.

It has been argued that adopting a qualitative approach reduces the level of reliability and validity of a study. When studying concepts such as technology adoption, leadership, culture, the digital divide, resources, and strategy, the predominant methods of study have been quantitative and reductionist.

An important observation regarding the two major concepts being researched (technology adoption and leadership) is that they both suffer from similar methodological constraints, i.e., major studies on technology adoption[24] and leadership[25] relied heavily on survey-based, quantitative approaches. These studies have been important in identifying key variables and enabling further exploration, but it may be argued that that methodology has been exhausted as a resource for new information. Consequently, a different approach may prove to be more fruitful in generating useful insights.

The Research Process

The research required a pilot study and two phases of data collection. The pilot study confirmed the suitability of questionnaires and other tools that were to be used in the first phase of data collection. It included references to all the relevant constructs, including culture, the digital divide, resources, etc. Ten respondents were interviewed to determine the suitability of the questions.

The first phase of data collection was aimed at meeting the following objectives:

[24] Bagozzi (2007), Fuchs et al. (2010).

[25] Choa (1998), Cole and Mehran (1998), Villalonga and Amit (2006), Welch (2003), Ghobadian and O'Regan (2006).

- To explore the relationship between leadership and technology adoption in owner-managed small firms.
- To look at the influence of internal factors such as strategy and resources on technology adoption in owner-managed small firms.
- To investigate issues such as the digital divide and culture to see how they affect technology adoption in owner-managed small firms.

The objectives listed above are three of the five objectives of this study, and they are all *output related*. This means that these objectives are expected to be met through the findings of the research itself as opposed to input/process related objectives, which aim to influence the activity of the data collection itself. In this study, four of the objectives are output related, and one is input related:

- To examine those factors and drivers that determine the different levels of technology adoption in travel agencies.

This input/process related objective emerged from the existing literature and guided the design of the instrument used in the first phase of data collection.

The instrument was constructed by using concepts that were tested in various studies related to technology adoption. In order to effectively probe and determine the appropriate questions to ask, concept maps were employed. These concept maps, in addition to variables and constructs used in previous research, were used as guides in the process of inquiry. These emerged from the rudimentary conceptual framework we used (Fig. 1.1). While conceptual frameworks typically do not emerge this early

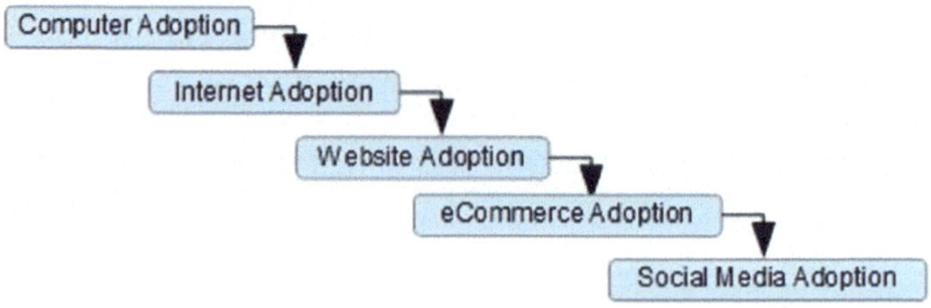

Fig. 1.1 Full stack of steps in the innovation interdependence perspective (*Source* Author's Creation)

in qualitative research, a rudimentary framework does help the researcher in streamlining ideas.[26]

As implied earlier, the interview instrument was a semi-structured one. This approach was chosen because the concepts under scrutiny, while not being new, are being applied in a different way. Despite the fact that a qualitative methodology was dominant in the research, the first set of questions on technology adoption used a closed-ended, quantitative approach ("yes" or "no" answers) in order to gauge their current level of engagement with technology.

The subsequent sections of the instrument were designed predominantly with open-ended questions, allowing for deeper explorations into attitudinal concerns and processes within the firm. These subsequent sections focused on organizational decision-making, leadership and a multitude of macro factors.

According to organizational decision-making theorists such as Huber and McDaniel (1986), organizational environments of the future are likely to be characterized by greater complexity. They argue that the decisions that are made within an organization, and *how* those decisions are made, are central to it and should be recognized as such. That being the case, it is important to get to the root of what drives those decisions, for both the organization and the individuals within it.

Given the complexity mentioned above, to have the instrument be guided by a single component model of attitude would be insufficient. The decision was therefore made to use the multicomponent model. The multicomponent model is an affective-cognitive model of attitude, which postulates that behavior may be most accurately predicted through a comprehensive understanding of the person, the situation and the interaction between person and situation. While this model has primarily been used in psychology and human behavior research, it is being applied to the organizational decision-making framework in this study. Notably, this is the first time that this paradigm is being applied to the study of technology adoption in business.

Rogers' (1994) *innovations adoption model* provides us with tools to assess leaders' perceptions of innovations. Given that the leaders of these firms are also the owners, the decision was made to focus on three characteristics that directly affected an owner's return on investment:

[26] Vaughan (2008).

- relative advantage
- perceived risk
- image.

Issues of Compatibility and Complexity were later assessed based on employee reactions to technologies.

- *Relative Advantage* was investigated using five statements that initially addressed the importance of technology in business processes and then moved on to more detailed questions around making work more efficient.[27]
- *Perceived Risk*, with regard to technology, was addressed in three open-ended questions. These questions inquired into the perceived risk of investment from the perspective of time and money.[28]
- *Image* was measured using two questions: how competitors view the firm and how the firm's own customers view it.[29]
- *Complexity* was determined through questions regarding how easy the technology was to use,[30] how easy it was to *learn* to use, and whether it actually made work easier or harder.
- *Compatibility* was assessed by how appropriate a technology was for the tasks being carried out in the organization.[31] Of note is a particular question, developed by the researcher, that asked how technologies were assessed before being introduced to the organization.

The sample for this study was comprised of the top executives of 31 travel agencies in Jamaica. While total agencies amount to 43, only those with owner-managers were included in this research. This was decided because they would clearly have greater levels of autonomy within their firms. Additionally, only those firms that were certified by the International Air Transport Association would be included and interviewed. The other firms were either a franchise or not certified by the IATA.

[27] Eason (1988).

[28] Wernerfelt (1984), Mowery et al. (1998).

[29] Grönroos (1993).

[30] Davis (1989).

[31] Peteraf (1993).

All 31 executives were interviewed as the only constraint was time. The questions were guided by the conceptual framework, but interviewees were allowed to speak freely. We started off the interview with closed-ended questions and proceeded to ask several open-ended questions while ensuring that all interviews were recorded and transcribed. The decision was made to comb through the data in order to identify any themes. This involved writing a case study based on all the data from the participating firms and then conducting a cross-case and comparative analysis.

This led to an exhaustive review of the responses to interview questions, which eventually revealed several common themes. The themes were shared among firms at similar adoption levels. Analysis of phase one data yielded several categories of technology adoption:

- ***Computer Adopters*** used computers but only for back-office accounting and front of house sales systems. Firms that use GDSs as their ONLY online sales tools are also included here. Why? Because the GDSs' suppliers provided them for free.
- ***Internet Adopters*** are those firms that use the internet for communication purposes (email, instant messaging, Skype, etc.).
- ***Website Adopters*** refers to firms that have company websites that are a part of their marketing and information strategy.
- ***E-Commerce Adopters*** use their websites for booking and payment facilitation.
- ***Social Media Adopters*** are actively participating in social media as a central part of their marketing, information and sales strategy.

Influenced by the work of Damanpour (1991), each firm was categorized in one of two states within each level:

- Initiation
- Implementation.

While analysis of the data from phase one provided us with the "what" (descriptions of the firms), it did not sufficiently tell us the "why." Why were each of the firms at their respective levels? Issues such as the digital divide, culture, strategy, and resources did not adequately answer the question as we anticipated they would, but leadership did appear to have some significance.

Now, even though there appeared to be some degree of causation when it came to leadership, it still could not explain each level of adoption that was observed; hence, it was not sufficient to build out a model. It was at this point that it was decided that a second phase of data collection would be necessary. This second phase would be guided by constructs in the most dominant leadership research at the time and aim to identify variables in leadership that influenced the level of technology adoption in firms.

The feelings of Owners were explored through in-depth, semi-structured interviews, which allowed for the application of Template Analysis.[32] This analysis allowed us to find common themes in the various narratives, enabling the inclusion of broader underlying structures in the analysis. This analysis was reinforced by utilizing a cross-case analysis matrix (cross-case analysis is a research method that can mobilize knowledge from individual case studies) and, since the interviews were recorded, respondents' own words could be used as an aid to more accurate categorization.

The objective of Phase 2 was to develop a model of staged technology adoption for owner-managed small travel firms that identifies the characteristics of leaders and firms at varying levels of technology adoption.

To meet this new objective, the second instrument was designed in a more focused manner but relied solely on open-ended questions. Given that leadership type was the major influencing factor of technology adoption for the travel firms, the second instrument's questions were grounded in the transformational leadership literature. The purpose being to strengthen and validate the leadership constructs and variables identified in phase 1.

This instrument uses, as a guide, the four constructs identified by Bass and Avolio (2003) in the Multifactor Leadership questionnaire. These constructs are:

- Idealized influence
- Inspirational motivation
- Intellectual stimulation
- Individualised consideration.

[32] Crabtree and Miller (1999).

The initial set of questions in the instrument attempted to clarify issues related to people's experience with technology, risk, and education. Each of these issues had emerged as a major theme during the analysis of phase one data, requiring deeper investigation.

All 31 owner-managed firms were invited for a second round of interviews, but only 20 of them agreed to participate. The major reason given by those firms that did not agree to the second round was time. Despite this setback, the number of firms that were willing to participate was sufficient for the study and a significant volume of data was gathered.

In this second phase of interviews, a relationship was identified between transformational traits and firm adoption levels. Those leaders who possessed more transformational characteristics shared commonalities such as:

- education level and type
- previous work experience
- family composition
- technology experience
- risk aversion
- and intellectual stimulation.

With this information, leadership typologies were developed for each level of technology adoption within the firms. While education levels of respondents were determined from revisiting phase one data, the remaining variables were gleaned from data collected in phase two.

Initially, out of all the variables, intellectual stimulation was the only one that seemed to influence technology adoption behavior for leaders but, once the other variables were combined in the analysis, greater synchronicities were uncovered. The template analysis technique was once again employed, along with a cross-case analysis matrix[33] and the results led to the generation of useful descriptive summaries. These summaries provided the tools to explain particular variables and led to an evaluation of how gaps in the literature may be filled.

Combining the data from both phases of the study conveniently allowed for the development of case studies. The approach taken was to examine individual cases and compare them, looking for similarities.

[33] Patton (2002).

This enabled the comparison of cases, and leadership characteristics, for firms at opposite ends of the technology adoption hierarchy.

Limitations and Impediments

This book provides for greater depth by viewing its research subject through the lens of *Critical Social Science*. It does this in two ways:

1. By giving credit to previously unearthed variables in the predominantly positivist studies of the last few decades.
2. By using the constructs and variables generated by those studies to probe deeper within the context of this research.

This allows for a deeper look into technology adoption and leadership while adding greater breadth to areas such as the resource-based view and strategic management.

Nonetheless, a major (and consistent) challenge was the time-constraint of owner-managers. Due, of course, to their integral involvement in the daily operations of the business. As a result, most interviews had to take place in the evenings when business activity was reduced. This had the knock-on effect of interviewees being tired during the hour-long interview. Apart from fatigue, interviewees were also concerned about being recorded. Fortunately, assurances that their anonymity would be maintained, by way of signed agreements, were sufficient to allay their anxieties. Said anonymity was practiced throughout the extent of the study and still remains in effect.

The Structure of the Book

Chapter 1 provides a review of the theories surrounding the decision-making process with regard to technology adoption. It addresses issues such as strategic management, organizational decision-making and technology adoption processes. More specific to technology adoption, the chapter delves into issues of the digital divide, culture, leadership and resources. The chapter concludes with two conceptual frameworks that illustrate how the aforementioned constructs interact.

Chapter 2 characterizes the firms that were interviewed and highlights both their current technology activities and their intention to use other technologies. This chapter also presents data about the role of strategy and resources when it comes to making decisions about adopting "new" technologies. Issues of risk are also addressed here.

Chapter 3 takes a closer look at macro factors like culture and the digital divide. How do intangibles such as norms, values, traditions and access to technology drive adoption? This chapter focuses on how external factors influence innovation and the making of leaders in firms.

Chapter 4 emphasizes leadership as the key determinant in technology adoption. While other factors do play a role, the data reveals that leadership is the most central. A discussion of the findings that reinforce this point is elaborated on in this chapter.

Chapter 5 provides a breakdown of how the emergent model from this research was developed. It begins with a reassessment of the framework presented in Chapter one and expounds on a revised iteration. Demonstrations are provided of how each component of the final model was identified and how they interact to provide a coherent model.

The Conclusion shows how this work meets the research objectives and contributes to the body of knowledge. It highlights the novelty of the research and its intended impact on theory and practice.

References

Amit, R., & Zott, C. 2001. "Value Creation in E-business." *Strategic Management Journal* 22(6–7): 493–520.

Bagozzi, R.P. 2007. "The Legacy of the Technology Acceptance Model and a Proposal for a Paradigm Shift." *Journal of the Association for Information Systems* 8(4): 3.

Bass, B.M., & Avolio, B.J. 2003. MLQ Multifactor Leadership Questionnaire. Redwood City: Mind Garden. www.mindgarden.com.

Buhalis, Dimitrios. 2002. *eTourism: Information Technologies for Strategic Tourism Management*. Essex: Pearson Education.

Choa, Myeong-Hyeon. 1998. "Ownership Structure, Investment and the Corporate Value: An Empirical Analysis." *Journal of Financial Economics* 47(1): 103–121.

Cole, Rebel A., and Hamid Mehran. 1998. "The Effect of Changes in Ownership Structure on Performance: Evidence from the Thrift Industry." *Journal of Financial Economics* 50(3): 291–317.

Crabtree, Benjamin, and William Miller. 1999. "A Template Approach to Text Analysis: Developing and Using Codebooks." In *Doing Qualitative Research*, edited by Benjamin Crabtree, and William Miller, 163–177. Newbury Park, CA: Sage.

Damanpour, Fariborz. 1991. "Organizational Innovation: A Meta-analysis of Effects of Determinants and Moderators." *Academy of Management Journal* 34(3): 555–590.

Davis, Fred D. 1989. "Perceived Ease of Use, and User Acceptance of Information Technology." *MIS Quarterly* 13(3): 319–339.

Drucker, Peter. 1990. *Post Capitalist Society*. New York: Harper Business.

Eason, Ken. 1988. *Information Technology and Organizational Change*. London: Taylor & Francis.

Fuchs, Matthias, Wolfram Hopken, Andreas Foger, and Martin Kunz. 2010. "E-Business Readiness, Intensity, and Impact: An Austrian Destination Management Organization Study." *Journal of Travel Research* 49(2): 165–178.

Gatty, Bob. 1998. "E-Commerce Gives Lodging Industry Edge." *Hotel and Motel Management* 213.

Ghobadian, Abby, and Nicholas O'Regan. 2006. "The Impact of Ownership on Small Firm Behaviour and Performance." *International Small Business Journal* 24(6): 555–584.

Ghosh, Shikhar. 1998. "Making Business Sense of the Internet." *Harvard Business Review* 124–135.

Golafshani, Nahid. 2003. "Understanding Reliability and Validity in Qualitative Research." *The Qualitative Report* 8(4): 597–607.

Govers, Robert F., Frank M. Go, and Kumar Kuldeep. 2007. "Promoting Tourism Destination Image." *Journal of Travel Research* 46(1): 15–23.

Grieger, M. 2003. "Electronic Marketplaces: A Literature Review and a Call for Supply Chain Management Research." *European Journal of Operational Research* 144(2): 280–294.

Grönroos, C. 1993. "An applied Service Marketing Theory." *European Journal of Marketing* 16(7): 30–34.

Huber, George, and Reuben McDaniel. 1986. "The Decision-making Paradigm of Organizational Design." *Management Science* 32(5): 572–589.

Jamaica Gleaner. 2011. "Tourism Minister Targets Illegal Travel Agents." Accessed July 27, 2011. http://jamaica-gleaner.com/gleaner/20110724/business/business7.html.

Kalakoa, Ravi, and Andrew B. Whinston. 1996. *Frontiers of eCommerce*. Reading: Addison-Wesley.

Klein, Stefan. 1996. *Interorganisationssysteme und Unternehmensnetzwerke*. St. Gallen: Universität St. Gallen.

Law, Rob, and Rita Leung. 2000. "A Study of Airlines' Online Reservation Services on the Internet." *Journal of Travel Research* 39: 202.

Lin, Pei-Jung, Eleri Jones, and Sheena Westwood. 2009. "Perceived Risk and Risk-relievers in Online Travel Purchase Intentions." *Journal of Hospitality Marketing and Management* 18(8): 782–810.

Minghetti, Valeria, and Dimitrios Buhalis. 2010. "Digital Divide in Tourism." *Journal of Travel Research* 49(3): 267–281.

Moital, Miguel, Roger Vaughan, and Jonathan Edwards. 2009. "Using Involvement for Segmenting the Adoption of E-commerce in Travel." *Service Industries Journal* 29(5): 723–739.

Mowery, David C., Joanne E. Oxley, and Brian Silverman. 1998. "Technological Overlap and Interfirm Cooperation: Implications for the Resource-Based View of the Firm." *Research Policy* 27: 507–523.

Neuman, W. Lawrence. 2006. *Social Research Methods: Qualitative and Quantitative Approaches*. 6th ed. Boston: Allyn and Bacon.

Patton, Michael Quinn. 2002. *Qualitative Evaluation and Research Methods*. 3rd ed. Thousand Oaks, CA: Sage.

Peteraf, Margaret A. 1993. "The Cornerstones of Competitive Advantage: A Resource-based View." *Strategic Management Journal* 14: 179–191.

Poon, Auliana. 2001. "The Future of Travel Agents." *Travel & Tourism Analyst* 3: 57–80.

Porter, Michael. 2001. "Bewährte Strategien Werden mit dem Internet Noch Wirksamer." *Harvard Business Manager* 5: 64–81.

Rogers, Everett. 1962. *Diffusion of Innovations*. New York: The Free Press.

———. 1994. *A History of Communication Study—A Biographical Approach*. New York: The Free Press.

Timmers, Paul. 1998. "Business Models for Electronic Markets." *Electronic Markets* 8: 3–8.

Vaughan, Roger. 2008. "Conceptual Framework." Power Point Presentation, Bournemouth University. Accessed May 14, 2010. http://www.powershow.com/view/a57f7-NDM1Y/Conceptual_Framework_Professor_Roger_Vaughan_May_29th_2008_flash_ppt_presentation.

Villalonga, B., and R. Amit. 2006. How Do Family Ownership, Control and Management Affect Firm Value? *Journal of Financial Economics* 80(2): 385–417.

Welch, Emma. 2003. "The Relationship between Ownership Structure and Performance in Listed Australian Companies." *Australian Journal of Management* 28(3): 287–306.

Wernerfelt, Birger. 1984. "A Resource-based View of the Firm." *Strategic Management Journal* 5: 171–180.

Werthner, Hannes. 1995. *Design Principles of Tourism Information Systems*. Wien, Oldenburg.

———. 1996. *Design Principles of Tourism Information Systems*. Innsbruck: Springer Verlag.

Werthner, Hannes, and Stefan Klein. 1999. *Information Technology and Tourism: A Challenging Relationship*. Vienna: Springer Verlag.
Williams, Densil, and Andrew Spencer. 2010 "Advertising and Tourist Arrivals: Evidence from Jamaica." *Tourism and Hospitality Research* 10(4): 359–366.
Wirtz, Bernd W. 2001. *Electronic Business*. Wiesbaden: Gabler Verlag.
Whyte, Millicent. 1983. *A Short History of Education in Jamaica*. London: Hodder and Stoughton.

CHAPTER 2

Leadership and Technology: Understanding Adoption Practices

Strategic management theory addresses topics such as competitiveness and the factors that drive it, and the long-term vision of organizations. One of the objectives of this study is to investigate the role of internal firm factors. The literature on strategic management provides scope for this as it relates to strategy and resources in the adoption of technology. The competitive approach, pioneered by Porter,[1] states that a company develops its business strategies to obtain a competitive advantage over its competitors.

In his seminal work, Porter points to several distinct competitive forces; however, much of the research has indicated that firm strategy must be influenced by not just competitors, but markets. While these are useful arguments, they neglect inputs such as ownership, leadership, and resources. This is a gap that this work seeks to fill. Also, at issue is the fact that studies that look at the impact of strategy on technology adoption (and other processes) have been merely superficial. Competitive strategy, therefore, must become fundamental parts of the adoption discourse.

In assessing the five forces model,[2] in conjunction with arguments posited by David (2007), there appears to be consistency surrounding the issue of competitiveness. That is, special attention must be given to

[1] Porter (1985, 2001).

[2] Porter (1985).

A. Spencer, *Technology Adoption in the Caribbean Tourism Industry*,
https://doi.org/10.1007/978-3-030-61584-0_2

both the external environment and the internal capabilities of organizations when it comes to survival within the same environment. As Porter stated, a company develops its business strategies by responding to five primary forces:

- The threat of new entrants
- Rivalry among existing firms within an industry
- The threat of substitute products/services
- Power of suppliers
- Power of buyers.

For companies operating in a globalized world, Porter's model of the five competitive forces addresses some key concerns. Companies must deal with these forces effectively in order to achieve a competitive advantage in the business environment. This model focuses on the value chains within an entire industry, not merely the mechanisms within a specific company.

It describes a network of interlinked players that perform value-creating activities. These other players must be considered in the strategy development for a company. As such, adding Porter's model to an e-commerce strategy can provide a useful framework for making decisions.

The work of Enright and Newton, however, did address a gap in Porter's studies. They asserted that the importance of competitiveness attributes might vary across locations, depending on product mix and target market segments. This is especially the case in complex, multifaceted industries like tourism.

A dominant argument from Porter's five forces model is that companies remain competitive by continually adapting to internal and external changes, and this is regardless of industry.[3] In fact, competitive strategies tend to influence other business strategies within firms, such as the distribution of products and services.

It seems clear, therefore, that where the external environment changes, companies must have the ability to keep pace to stay ahead of the competition. If they hope to compete, it cannot be business as usual.

Christopher[4] defines distribution as "the process of strategically managing the movement and storage of materials, parts and finished

[3] David (2007).

[4] Christopher (1992, p. 4).

inventory from suppliers through the firm and onto its customers." He argues further that distribution strategies involve a more comprehensive consideration of logistics, including inventory levels, materials management and information systems (as it relates to transport). While there was an awareness of the need for competitive distribution strategies, there was also the need for new management skills. These new management skills were required to facilitate the implementation of said competitive distribution strategies.[5] These new management skills include:

- systems management
- customer service management
- operations coordination.

The literature on distribution strategies, however, shifted focus from internal management of the organization to emphasize on channels strategy. Gattorna (1994) states that the most significant strategical challenge facing organizations is the development of a channels strategy. While much of the earlier literature focused on the distribution of tangible product, Gattorna highlights that any progressive organization, including those that trade in intangibles, should attempt to collect the right strategy to get the final product or service to the consumer.[6]

Despite this, his definition of distribution channels still reflects a preoccupation with tangible goods. He defines the distribution channel as a commercial arrangement to transport a product from the point of production to the point of consumption. It has been identified here that, even with a focus on tangible distribution, sometimes a change of company would be needed or merely appropriate.

In order to establish a distribution channel, three determinants are required.[7]

1. *The requirements of the final customer, as measured by an aggregation of customers with similar requirements.* The channels strategy will depend on the potential customers in the segment and what route

[5] DeHayes and Taylor (1972), Persson (1978), Christopher (1986).

[6] Chandler (2003), DeHayes and Taylor (1972), Persson (1978), Christopher (1986).

[7] Rushton and Oxley (1993), Gattorna (1994), Hatton (1994), Chorn (1994).

will reach them most efficiently. In the context of e-commerce adoption, this may be determined by whether or not the market segment is predisposed to online transactions.

2. *The capabilities of the organization.* The resource-based perspective says that a firm's strategic decisions are, in large part, influenced by its resources, which may also allude to management capabilities.
3. *The availability and willingness of intermediaries to participate in the channel.* The discussion of intermediaries has focused on their role as a conduit for the supplier; however, very little is said about the determinants of their distribution strategies.

Discussions of channel strategy, however, moved to a broader debate of managing the entire supply chain. Within the discourse on international distribution,[8] it is posited that companies involved in said industry must be carefully managed to ensure that goods and services reach their target market efficiently.[9] This shows an emphasis on the needs of the consumer and for the supplier to use their resources efficiently. It puts the responsibility of controlling these factors, including the intermediaries, on the supplier and ignores the fact that the intermediary itself must also manage their bidirectional relationships within the supply chain.

Schary and Larsen (1998) state that the concept of the supply chain embraces several elements. Firstly, it identifies the complete process of providing goods and services to the end-user/consumer and brings the activities of supplier and customer into a single system. This is coordinated through an information system that is accessible to all members. They highlight that while the main objective of the supply chain is to service customers, this must be balanced against costs and assets.

A chain is only as strong as its weakest link; hence, members of the supply chain do not do this on their own. The achievement of efficient performance of the supply chain is dependent on the performance of the chain as a whole. And this requires the management of a network of interconnected businesses[10] to achieve the least total logistics cost.

The travel supply chain, like supply chains in many other industries, involves the supplier, wholesaler, retailer, and consumer. The suppliers

[8] Rushton and Oxley (1993).

[9] Chorn (1994).

[10] Harland (1996).

are those that provide services—airlines, hotels, attractions, and car rental companies. Wholesalers refer to companies such as tour operators and companies, like Thomas Cook and Thomson Travel, that create travel packages. Finally, at the end of the chain, retailers are travel agencies that sell these tour bookings and travel packages to the consumers. Being the contact point to the consumer, it is vital to investigate what factors drive or deter the travel agents' engagement in e-commerce.

Hamel and Prahalad (1994) observed that, in the effort to secure a more significant global market share, an effectively managed distribution chain is the most crucial element in a global strategy. Effective management will involve managing costs and satisfying existing market share while claiming new market share. Travel firms must apply these principles while meeting the competing needs of suppliers and consumers.

It costs to establish a presence within markets, so suppliers must conduct cost–benefit analyses to ensure that the benefits of gaining access to said markets are worth the costs. Many firms, such as Google, do this through strategies that enable them to give away their product for free. Terpstra (1991) adds that, from a marketing perspective, there are three distribution tasks:

- gaining entry into a foreign market
- serving multiple markets simultaneously
- establishing a presence as an insider to each market.

Supply chains can be manipulated and modified depending on the needs of the supplier and customer. Managers who are alert to the realities of the marketplace are constantly seeking a competitive advantage. One cannot assume that good products will sell themselves, nor that today's success will carry forward into tomorrow.[11] It is on this basis that organizations are constantly seeking new sources of revenue while minimizing distribution costs.

While contemporary theorists focus on the distribution strategies of suppliers, it may be argued that keen attention should also be paid to the intermediaries within the chain. These firms must also make critical decisions regarding how to reach their markets and transact business. Given

[11] Christopher (1998).

the reduced transaction costs and commissions, travel intermediaries should be eliminated entirely.[12]

When Caribbean travel agencies attempted to boycott the sale of American Airlines tickets in 2000, their efforts proved futile. To further the point, Hatton (2004) points out that the reduction of commissions paid and the competition provided by the internet present an uncertain operating environment. Further still, in 2003, British Airways reduced agent's commission in the United Kingdom to 1%. What is clear from this trend is that organizations must do whatever is necessary to ensure profitability. Either by increasing revenues, reducing costs or both. With inflation, competition, and rising oil prices, revenue increases may only be marginal as lower prices may not be passed on to the customer. Given these issues, the main focus is, therefore on reducing costs, especially those related to distribution.

Pearce (2009) argues that suppliers and their location destinations must seek to develop more effective distribution strategies. He contends that, in trying to understanding distribution design, the existing tourism-marketing literature has taken a tier-by-tier approach rather than a more useful network approach. This, he argues, has led to channel width considerations being poorly addressed.

However, while the network approach is useful, it fails to address the deeper operational concerns of specific players. And for those cases of research (on tourism) that do pay attention to specific players, it tends to be on either suppliers or consumers. Alternatively, the literature gives attention to the relationship between suppliers and wholesalers[13] or between intermediaries.[14] This work contends that greater emphasis should be placed on *why* travel retailers choose the business strategies that they use.

There has been much study done on travel agencies,[15] but this work is focused on what agencies need to do to avert disintermediation—namely the issue of internet adoption. Despite compelling arguments, many firms

[12] Buhalis and Licata (2002).

[13] Crotts et al. (1998), Karamustafa (2000).

[14] Lumsdon and Swift (1999).

[15] Palmer and McCole (1999), Barnett and Standing (2001), Dale (2003), Law et al. (2004).

have yet to adopt the internet as a part of their marketing and sales strategy. It is the objective of this work to determine why that is.

Intermediaries do not operate at either end of the supply chain. Therefore, an investigation into their relationship with suppliers is critical to understanding the strategies which they employ. The concept of strategic purchasing is not a new one,[16] and neither is the concept of relationships in business. Theorists have suggested that business buyers should create a strategy that consists of four elements:

- self-analysis
- vendor analysis
- vendor performance
- program review.

The purchaser must conduct detailed research before entering into an agreement. It is vital that buyers adopt a systematic approach to transacting with vendors, because the stock you by, your inputs, will invariably affect your outputs.

A major element in the field of strategic procurement management is the ability to manage relationships with other firms in the supply chain.[17] According to Cox (1998), an effective business strategy recognizes that a firm's boundaries must change in response to its consumers' preferences. This theoretical position was built on the works of other theorists[18] who focused primarily on reactive and simplistic approaches to purchasing and supply management. Most of this previous analysis assumed that a firm's operating structure would remain fixed, and so focused on the interplay between asset specificity and transaction costs.

Despite this, the issue of asset specificity was treated poorly in the literature. This is because, due to the nature of their operations, most intermediaries aren't capable of sourcing internally and so must source externally, regardless of whether services have low or high asset specificity. Cox, therefore, extends the discussion to address shifting boundaries and, what he calls, "strategic supplier alliances." These are alliances, often also

[16] Eames and Norkus (1988).

[17] Cox (1998).

[18] Ricketts (1994), Williamson (1979, 1990), Reve (1990).

referred to as "joint ventures," where very close, single-sourced relationships may emerge between the main contracting firm and a supplier of a complementary product or service.

The systems used by the primary supplier may influence those used by the complementary supplier. In the case of an intermediary that relies on a sole supplier for stock, this influence is likely to be more pronounced. For example, if the primary supplier only sells stock online, then the secondary or complementary suppliers may be somewhat "coerced" into implementing systems that facilitate the purchase of stock online. This is especially the case where relationships like those discussed by Cox are manifested. It is on these grounds that he argues for a relational competence analysis. Relational competence analysis involves a determination of what the efficient boundaries of the firm are so that these new boundaries can be set to reduce transaction costs and improve quality and value, whether collaboratively or competitively.

Virolainen (1998) however, describes these approaches as "narrow," since the literature focuses on particular products but less so on the actual procurement function. Instead, he proposes an integrated procurement strategy that involves the following:

- value-chain positioning
- objectives setting
- organization structuring
- strategic make-or-outsource decisions
- choice of different strategies
- company-level strategy
- functional strategies.

However, while Virolainen has attempted to produce a more general framework, what is needed is even greater specificity as the nature of procurement varies by industry and relationship. Additionally, while studies have focused on the acquisition of services, little has been done to assess the nuances of procuring a product.

According to Ryals and Humphries (2007), there has been a quiet revolution in supply-chain management. Whereas before, the focus used to be on sourcing least-cost transactions, there has recently been a shift in emphasis toward the formation of long-term relationships with a small number of key suppliers. This is in contrast to the former prevailing view that transaction cost is the key driver.[19]

This further reinforces the point made earlier that, through strong relationships, systems used by primary suppliers may influence the systems used by complementary suppliers and intermediaries.

To date, procurement strategies focus primarily on offline modes of distribution. However, given that, in most cases, supply-chain related activities account for the spending of 70% of a firm's revenue,[20] then it is recommended that other options, including e-procurement, be explored.

A discussion of e-commerce is vital as it looks at online transactions in both directions along the supply chain:

- *e-procurement*, for business-to-business transactions
- *e-distribution*, for business-to-customer transactions.

These considerations have significant implications for firms' business strategies.

ICT can change the structure of an industry and alter the rules of competition.[21] It can also be used to create a sustainable competitive advantage for companies, by providing them with new competitive instruments, and allow them to develop new business within their existing activities. Previous research has been preoccupied with how ICT modifies strategy and industries but has failed to demonstrate how different types of strategy affects levels of technology adoption. Recently, however, Beckinsale et al. (2011) have emphasized the role of strategy in ICT adoption, and it is contended that this reconceptualization is of critical importance. The emphasis is, therefore, on each decision-maker's performance.

Meanwhile, those researchers that have based their work on broader, organization-based decision-making, have built their work on the foundations set in earlier work by Cyert and March (1963).

In an attempt to broaden the scope of the discourse, Langley et al. (1995) recommend that future research into organizational decision-making must address more complex issues. Issues such as the fact that decisions are not made purely by cerebral rationality, but also by the

[19] Ricketts (1994), Williamson (1979, 1990), Reve (1990).

[20] Presutti (2003).

[21] Porter and Millar (1985), Porter (2001).

experiences of the decision-maker. Additionally, single decisions should be viewed as the result of the interaction of multiple issues. This suggests dynamic linkages rather than simple linear ones.

Kwon et al. (2009) advocate for an integration of both approaches in order to assess micro decision-making within the context of macro or organizational decision-making. With this in mind, this book attempts to identify those concepts that have been identified as having an impact on technology adoption, and to observe how those concepts interact in organizational decision-making more keenly.

Hubert and McDaniel emphasize that organizational environments of the future will become more complex. As such, they suggest that firms recognize the centrality of their decision-making mechanisms, and get to the root of what drives decisions for both individuals and organizations.

To this end, Langley et al. (1995) emphasize those individuals who are the key decision-makers within firms.

This work seeks to inform the theoretical analysis by further analysing those general typologies that have already been recognized within existing decision-making frameworks. Typologies such as:

- Decision-maker as *creator*—an intuitive individual who drives organizational decisions through creative insight.
- Decision-maker as *actor*—an individual who passively acts in accordance with the current context that an organization finds itself. Such as problems or opportunities.
- Decision-maker as *carrier*—an individual who carries their experiences with them and is guided by the impact of their environment.

Previously, far less emphasis was placed on the vision and abilities of the decision-maker, with the seminal literature[22] considering them as more of a "receptacle" and thus treating the individual as an *actor*. More contemporary research,[23] however, declares the importance of insight in the role of decision-makers and, thus, treats the individual as a *creator*. Of the three recognized typologies, however, the least emphasized is that of the *carrier*. This may be explained by the fact that both *creators* and *actors*

[22] Simon (1957), Cyert and March (1963).

[23] Hitt et al. (2001), Ghobadian and O'Regan (2006), Falk (2008).

naturally have experiences based on local and global context. This is still accounted for in the literature on culture and the digital divide.

Before continuing, it must be stated that decision-makers are often referred to as leaders,[24] so the two terms may be used interchangeably throughout this text.

A significant development in the literature on organizational decision-making came from an assessment of leadership styles. This is the distinction between transformational and transactional leadership.[25] These two forms of leadership have been shown to have distinct different impacts on decision-making styles.

A transformational leader is one who is known to be a visionary—a decision-maker who is charismatic and influences strategic transformation through motivation.[26] A transactional leader, on the other hand, is defined as an efficient manager who focuses on the task at hand and uses a reward-driven approach.[27] These types of leadership affect whether a leader adopts a restricted or comprehensive decision-making style.[28] This work will assess the leadership role in the process of decision-making by analyzing these typologies within various situational context.

There has also been some focus on what is referred to as ***pooled linkages*** and ***contextual linkages***. Pooled linkages are those decision-making issues that are linked through competition. This could be competition for financial resources, managerial time and energy or any other limited resource. Contextual linkages refer to situations in which the individual and/or organization makes decisions within a culture, ideology, structure, and/or strategy context.

Analyzing through a Critical Social Science lens allows the various phenomena under scrutiny to be observed subjectively while acknowledging objective limitations. This is a feature of the aforementioned ***bounded autonomy*** concept. The discussion of the conceptual framework at play begins with a critical review of the literature on technology adoption models. More specifically, the behavior being focused on within

[24] Victorino et al. (2006).

[25] Tatum et al. (2003).

[26] Kuhnert and Lewis (1987).

[27] Ibid.

[28] Tatum et al. (2003).

that model is the process of internet adoption, particularly for sales and marketing purposes.

In light of limited resources, the decision to adopt technology-based innovations must be a calculated one. Hence, understanding the considerations that are involved in calculating whether or not to adopt new technology is crucial. In some cases, decision-making may be a complex process, involving multiple parties to approve adopting new technology, while in others it may be a relatively simple unilateral decision by a single leader.

Diffusion and Adoption

Diffusion and adoption are two distinct but relevant concepts in this discussion. Whereas adoption is making use of an innovation within operational practices and processes, diffusion refers to how information about an innovation is transmitted in a group.[29] This work contends that both of these concepts are interrelated and symbiotic. Knowledge transmission (diffusion) may affect the decision to use (adoption) and, inversely, adoption will both necessitate and facilitate diffusion.

The diffusion and adoption paradigm saw significant growth during the 1940s,[30] but two decades later output from those studies seemed to suffer from limited applicability due to the insularity of the specific field. This was because of the output, from that phase of research, was constrained and did not apply to other fields. In 1962, Rogers developed a more general theory that was able to apply the paradigm from earlier research to be useful across disciplines. This led to the wider application of the theory of diffusion and adoption of innovations to broader fields. Moital et al. (2009) even state that Rogers' contribution to the theory is so significant that his model "still prevails as the main theoretical source in the study of diffusion and adoption of innovation."

The work of Everett Rogers[31] has been foundational in the discussion on innovation diffusion. In *Diffusion of Innovations*, he explores why obviously beneficial ideas take so long to become adopted into as mainstream practice. Rogers (2003) defines *diffusion* as the process in

[29] Rogers (1994).

[30] Ryan and Gross (1943).

[31] Rogers (1962, 1976, 1983, 1994, 2003).

which an innovation is communicated through certain channels over time among the members of a social system. The key variable in this context is that *new* ideas are being communicated.

Rogers goes on to list the key elements of technology diffusion as:

- Innovation
- Communication channels
- Time
- Social system.

Rogers (2003) refers to innovation, distinct from *technology*, as any new idea, practice or object. The critical variable, as aforementioned, being its "newness." While this definition may include technology, it does not necessarily refer to it nor include it as innovation is much broader in scope.

While earlier approaches simply identify innovations as positive,[32] Latzer (2009) looks at the possibility of those innovations being disruptive or unsustainable. Keller (2008) meanwhile, states that no single factor can determine the success or failure of an innovation, but that an innovation's effectiveness is dependent upon interacting factors. Factors that may be social, political, and/or economic in nature.

The term "communication channel," in the context of diffusion, encapsulates how a new idea passes from individual to individual[33] or through a group of people.[34] When considering these two concepts, diffusion of innovation and adoption of innovation, it must be borne in mind that diffusion is a critical prerequisite for adoption. One cannot adopt a new technology without first learning about it and being educated on its usefulness. It must also be borne in mind that while diffusion is critical to adoption it does not guarantee it.

Hoffman and Roman (1984) identified two key dimensions of information flow within organizations:

- the origin of the information
- The emphasis on innovation.

[32] Bagozzi (2007), Lederer et al. (1998).

[33] Rogers (1994).

[34] Loudon and Della Bitta (1993).

It should be pointed out that although leadership is not explicitly mentioned, it is inherent in their argument. Particularly so in the second dimension, although this dimension may also be influenced through organizational culture or strategy. Additionally, those innovations that are introduced via leadership are granted greater legitimacy. This work aims to emphasize the leadership role in the diffusion of innovation more clearly. For instance, while the flow of information is easier than ever before thanks to the internet, what drove the dissemination of information on the internet itself?

It is at this point that Roger's representation of social systems becomes of particular interest. Roger states that a social system is a set of interrelated units that are engaged in joint problem solving to accomplish a common goal. Within this definition, "units" refer to organizations or informal groups.

Roger's conceptualization is particularly useful as the scope of the "social system" is scalable. It applies to actors within a small village as well as to distinct groups within a country. While his work has been applied generally, Rogers suggests that context is important, and it is within the context of diffusion of technology, Rogers contends, that social structure and the norms of the society are important.

Yun and Avvari (2011) have contributed to the discourse by highlighting gaps in the research. These gaps are "filled" by emphasizing the role of "innovation actors." An example of an innovation actor in action would be leaders in the public and private sector who champion new ideas based on recent innovations. This further endorses the idea that the leadership element is core to innovation diffusion.

Roger's work (2003) indicates that opinion leadership may be very influential, and so leaders may act as change agents in the diffusion process. He continues by declaring some very useful distinctions in types of innovation decisions:

- Optional
- Collective
- Authority-driven.

Optional innovation decisions rarely result in comprehensive adoption by the group. *Collective innovation decisions*, while having gained more traction because of general consensus by the group, are largely leadership driven. In such instances, the leader is usually a charismatic individual[35] or a thought leader.[36] *Authority-driven innovation decisions* typically yield compliance, but other powerful individuals and groups may also be opposed to the diffusion of particular technologies. This highlights the importance of leadership/ownership in the discussion of innovation diffusion.

While collective and authority innovation decisions are more common, they are still leader-initiated. This highlights the need for greater focus on leaders as decision-makers. Brown (1981) goes on to present several perspectives on innovation adoption.

The *Economic History Perspective* claims that an innovations' chances of adoption are proportionate to its usability to the market in terms of value for money. The *Adoption Perspective*, pioneered by Rogers (1962) and the most dominant and mature perspective on the subject, holds that innovation adoption is the result of the communication process.

The *Market and Infrastructure Perspective*, on the other hand, seeks to move away from the individual approach, claiming that opportunities are "purposely unequal." Instead, it focuses on the conditions within the given context to determine how those conditions facilitate innovation diffusion.

The *Development Perspective* is an extension of the market and infrastructure perspective insofar as it attempts to identify the impact of innovation diffusion on the welfare of society. Unlike the former three perspectives, however, it challenges the assumption that innovation diffusion is always a net positive. Toward this end, Brown presents examples of developing nations that have experienced widening inequality and a strengthening of the elite class as a result of adopting new technologies.

While each of these perspectives is useful, the ones most applicable to contemporary analysis are the Adoption Perspective and the Market and Infrastructure Perspective. The two of these viewpoints emphasize both the individual as well as the broader environment. This works in tandem with the technology acceptance model.

It has recently become apparent that innovations are adapting to markets in ways that, seemingly, ignore geography. This flies in the face of theoretical frameworks presented by perspectives like the Economic

[35] Hitt et al. (2001), Bass et al. (2003).

[36] Rogers (2003).

History Perspective,[37] as geography used to be a prime factor in the spread and evolution of innovations, and the Market and Infrastructure Perspective, as different geographical locations have differing infrastructure.

The Development Perspective, on the other hand, promotes equal opportunities, but it does not explain why people do or do not adopt new technologies and/or ideas. The Adoption Process, outlined by Rogers (1994), answers this question.

In the Adoption Process, Rogers suggests a linear progression:

1. Knowledge
2. Persuasion
3. Decision
4. Implementation
5. Confirmation.

While this proposed progression has been useful in previous research, more recent research, including work done by Straub (2009), has found that the adoption process is not quite so linear nor so clearly defined. While Rogers has since granted that the process steps may be rearranged, he maintains that the process itself is still a linear one.

In contrast, Woiceshyn (2000) argues that, specifically, within organizations, adoption is a learning process. This learning process involves:

1. Observing
2. Interpreting
3. Integrating
4. Acting.

Despite this linear description, and unlike the steps in Rogers' Adoption Process, these steps may happen concurrently. Woiceshyn's work also recognizes several facilitators of the adoption process:

- Capability
- Resources
- Shared Values

[37] Kotler and Armstrong (2008).

- Effort
- Motivation
- External factors.

Straub (2009), meanwhile, considers adoption be a complex social and developmental process that may be influenced by several other factors, including individual perceptions, related emotions, and general context. From Straub's point of view, the relevance of the unique subjective experience is paramount. This is a view that has only been directly addressed in the innovation diffusion literature.

The work of this book aims to show that those factors that have been classically understood to influence technology adoption affect, and in turn are affected by, personal factors of the many individuals involved. In particular, those personal factors that apply to leaders. This revelation turns these elements from being direct factors to mediating variables. Yes, they facilitate adoption, but they are complementary to the process.

According to Zhou (2008), innovation diffusion research may be classified among three levels:

- Individual
- Organisation
- National.

While these categories may vary in scope, they do share many of the same drivers of innovation diffusion:

- Personal
- Situational
- Social
- Socioeconomic
- Market
- Infrastructural.

Zhou also argues that a fourth level should now be considered: Intra-organizational.

The Intra-organizational level emphasizes that if an organization adopts innovation, it will not be adopted uniformly throughout that organization. People and groups within the organization will respond differently to the adoption, depending on factors relevant to their situations. Zhou goes further, classifying the different levels of adoption-responses:

- Voluntary adopters
- Forced adopters
- Resistant non-adopters
- Dormant non-adopters.

This fourth level takes into account the willingness of individual members of the organization. How willing are they to personally adopt the new technology that has been formally adopted by their organization? The Inter-organizational level looks at individual responses to variables such as leadership, resource use and strategy while taking into account concepts such as the diffusion of innovation[38] and frameworks like the Technology Acceptance Model,[39] both of which are related to the Broader *Theory of Reasoned Action*.[40]

The Theory of Reasoned Action states that an individual's behavior is the result of the interaction between their attitude toward a specific activity and the wider society's view toward said activity. The theory argues that the core of innovation adoption lies in the adoption of new ideas by the individual within the context of the surrounding society. This is not to say that there have not been studies that look at personal choice within the context of the wider society; the diffusion of innovations theory does just that. What they do, however, fail to do is to look at that individual process from a leadership perspective. Ironically though, it is implicitly dealt with in the field.

Rogers' theory of diffusion of innovation was not the only one to consider adoption from an individual perspective. Davis' Technology Acceptance Model contributed advanced ideas related to those factors that act as adoption drivers for individuals. According to Davis' Technology Acceptance Model (TAM), the perception of a new technology's

[38] Rogers (2003).

[39] Davis (1989).

[40] Fishbein and Ajzen (1975).

perceived ease of use and perceived usefulness, by the individual, are major considerations and have a significant influence on that individual's acceptance of said technology. Lederer et al. (1998) also support this notion.

For context, perceived ease of use refers to how easy the technology will be to learn and to use. In contrast, perceived usefulness is the degree to which it is believed that the technology may enhance time and output efficiency of related work and tasks.[41]

The TAM is at the other end of the spectrum relative to most other research in this field and, likewise, suffers from similar weaknesses. It focuses on the cognitive processes of the individual, essentially making it a micro decision-making model. The fact that the TAM does not take broader organizational nor societal factors in mind makes it limited in its approach.

People's attitude toward technology also influences their perceptions of usefulness and ease of use of technology,[42] which ultimately affects their usage of said. This makes the concept of technology attitude critical to the discussion on technology adoption.

As said before, research like the TAM suffers from the same issue as previous research studies in that they only look at half the picture, either the micro or the macro. By focusing on personal attitudes and behaviors, they've neglected the impact of broader issues.

Arguably, personal and subjective factors like perceived usefulness, perceived ease of use, and attitudes to technology do not operate in a vacuum but are, in turn, affected by other objective elements such as one's age, education level and other demographic factors. Legris et al. (2003) expound on this point, arguing that, apart from these personal variables, other social influence factors must be considered for, as the saying goes, "no man is an island unto himself." As such, this study aims to bridge this gap and to take a broad view, encompassing both ends of the spectrum, the micro and macro, in analyzing technology adoption.

Before any technology can be used, one must be taught how to use it. That makes the role of *learning* paramount in the discussion of adoption. Bagozzi (2007) initiated a debate on the role of learning to use a computer, looking at the global adoption of that technology. Previous to

[41] Lederer et al. (1998).

[42] Wöber and Gretzel (2000).

this research, the focus on computer adoption discussion was limited to the usage of the technology. The research posits that, despite how well-built and well designed the computer system itself may be, inadequate learning of said systems can substantially affect adoption.

This also brings the beliefs, attitudes, and intentions of individuals into stark relief. The research had primarily focused on attitudes: attitudes to success, failure, and toward the process of pursuing goals. In other words, the research separated the "intention to try" from "actually trying," since it was the act of learning that separated intention to use from actual usage. For instance, an individual may display an "intention to try" to use a computer by actually purchasing one, but if there is no follow-up action in learning to use the computer ("actually trying"), then usage would naturally suffer. A person cannot use a computer the things they know how to do. If they have not learned anything about using a computer, then it becomes merely an expensive paper-weight.

The natural implication of these findings is that organizations, in adopting new technologies, must make learning the new technology a part of their adoption process. This means implementing strategies that motivate the organization's members to learn the new technology. This is a challenge, given the varying levels of personal attitudes toward learning and adoption.

This illustrates the importance of *compatibility*. Rogers defines compatibility as "the degree to which an innovation is perceived as being consistent with the existing values, past experiences and needs of potential adopters."

In contrast to TAM, the Unified Theory of Acceptance and Use of Technology Model identify that, in addition to personal factors like attitude to technology, the conditions that facilitate usage within the particular context and the influence of society also play a part in motivating usage.[43] Other theorists have also elaborated on situational influences and their impact on communication technology usage.[44] It has been proven that the social environment exerts influence on the perceived applicability of a technology.

If one observes others in their society using or refusing a particular technology, then that will affect that individual's perception of the

[43] Dulle and Minishi-Majanja (2011).

[44] van den Hooff et al. (2005).

applicability of that technology to their own needs. This discussion, recognizing social influences, extends the earlier work of Rogers' Diffusion of Innovation Theory (1983). Together, the work of Rogers (1962) and Davis (1989), identify contrasting influences and discuss their effect on perception across the spectrum, from the micro level (individual) to the macro level (organization).

The Diffusion of Innovations Theory looks at a broad set of influences while the TAM is tightly focused on the personal cognitive influences within individuals. This makes the Diffusion of Innovations Theory more appropriate for the goals of this study. Fuchs et al. (2009) contribute to the debate on technology adoption by pointing out that the availability of a suitable ICT infrastructure and the context of the surrounding environment are also key drivers.

It has been argued, however, that these considerations are insufficient insofar as they fail to explain why half of the newly implemented innovations within the United States have been considered as failures.[45] Lippert and Davis (2006) have suggested that "trust" is an element that should be considered and, hence, has introduced this into their conceptual model. They posit that "technology trust" and "interpersonal trust" have a positive effect on "planned change activities," enhancing the level of adoption. Despite these recent developments in the overall discussion, the main thrust of the debate still lies with firmly entrenched concepts laid out by formerly established theorists.[46]

The situational and personal influences mentioned earlier speak to many factors, including:

- Media exposure
- Compatibility
- Perceptions
- Social factors
- Socioeconomic factors.

[45] Lippert and Davis (2006).

[46] Rogers (1983), Davis (1989), Lederer et al. (1998), Wöber and Gretzel (2000), Legris et al. (2003), Vishwanath and Goldhaber (2003).

There has also been discussion regarding negative personal and situational influences that may adversely affect technology adoption and ways of mitigating against this. One such mediating factor is training.[47]

While the TAM has proven to be extremely useful in analyzing situational influences, it does have its drawbacks. Namely, it ignores some of the critical arguments in the established literature. In particular, disputes related to variables such as resources, strategy, ownership, and leadership. In order to conduct a more detailed investigation into why specific firms engage in technology adoption, while others do not (or not to a similar degree), a closer look must be taken at how these variables mentioned above interact.

According to Moital's (et al.) Innovation Interdependence concept, understanding what determines a firm's level of technology adoption is key to promoting higher levels of technology adoption, including e-commerce adoption, in other firms. One school of thought posits the case that technology adoption, more specifically, e-commerce adoption, is determined not by the firm, but by the firm's customers.[48]

They argue that some firms claim to engage in e-commerce and they do have a company website. Still, it is a static website that only provides information and is not built on (or with) a functional e-commerce platform that will provide customers with an interface to engage in transactions. Instances such as these may be categorized as *Staged Adoption*, a phenomenon that has been pointed out by Daniel et al. (2002).

Daniel et al. have gone further by categorizing those firms that are in the process of adoption:

- *Developers* are those firms that are in the early stages of making information about their products available online.
- *Communicators* use email and other messaging apps like Skype to communicate with customers.
- *Web Presence* designates those firms that have online booking and ordering facilities.
- *Transactors* maintain fully functional e-commerce platforms online where they actively sell their products and services and engage in information gathering.

[47] Bagozzi (2007), Marler et al. (2006).

[48] Zappala and Gray (2006).

Those firms that were initially established as physical, brick-and-mortar operations tend to not advance to the Transactor stage of adoption.[49]

While resource constraints are usually cited as blockers to this advancement, Gray (2006) posits that ICT is an excellent option for overcoming resource limitations, allowing businesses to grow beyond their perceived limits. Unfortunately, some small and medium-sized businesses, especially in developing regions such as the Caribbean, find it difficult to source the initial capital required to take this step, arguing that such expenditure would not provide a sufficient return on investment. This perspective may be more a matter of attitude and the general climate toward innovation than fact.[50]

Zappala and Sarchielli (2006) contrast the fact that, interestingly, while online selling has stagnated, online purchasing has risen. This illustrates the fact that e-commerce adoption has been on the decline.[51] Mochrie et al. (2006) argue that firms simply do not have the capacity, physical and/or human, to develop and implement effective adoption strategies. As it is, some firms may reject adopting innovation on that basis alone. In fact, according to Rogers,[52] a rejection of innovation, or discontinuance thereof, may occur at any stage of the adoption process.

Rogers also cites certain pre-existing conditions that may affect the adoption process, such as:

- Previous practice
- Felt needs/problems
- Innovativeness
- Social norms.

Each of these conditions plays a part in e-commerce adoption and particularly in the Caribbean.

Historically, this region proceeds relatively slowly with regard to innovation, due to the resource factors mentioned above, but also due to loyalty to traditional ways of doing things. Despite this, Tanpong et al.

[49] Lippert and Davis (2006).

[50] Zappala and Sarchielli (2006).

[51] Houghton and Winklehofer (2004).

[52] Rogers (1962, 1976, 1983).

(2009) highlight that, for digitally related sectors, e-commerce becomes a strategic imperative, while for non-digital sectors, it is a strategic choice.

Research conducted during the 1990s and the early 2000s created several categories of key drivers to adoption decisions:

- Customer pressures[53]
- Competitive pressures[54]
- Supplier pressures[55]
- Industry challenges[56]
- Company challenges.[57]

Collectively, all of these forces have been influenced by the worldwide presence of the internet.

Many theorists agree that the internet has fundamentally changed the way that established companies compete and do business.[58] It has done this in three ways:

- Information asymmetry reduction[59]
- Disintermediation
- Reintermediation.

For a customer to learn about the features and prices of a product, they no longer have to visit a physical store. They can simply look it up online. They don't have to waste the time and energy it would take to search for the information physically. Likewise, what this means for companies

[53] Poon and Joesph (2001), Bigne et al. (2008), Daniel et al. (2002), Dyerson and Harindranath (2007).

[54] Bigne et al. (2008), Patricia (2008), Teo et al. (2009), Werthner and Klein (1999).

[55] Beekhuyzen et al. (2005), Buhalis and Deimezi (2004), Vrana and Zafiropoulos (2006).

[56] Grandon and Pearson (2004), Kuan and Chau (2001), Saffu and Walker (2008), Teo et al. (2009).

[57] Bennett and Lai (2005), Bigne et al. (2008), Heung (2003), Law et al. (2004), Stansfield and Grant (2003), Warden and Tunzelana (2004).

[58] Rayport (1995), Choi and Stahl (1997), Grieger (2003), Williamson and Scott (1999), Afuah and Tucci (2003), Gentner et al. (2017), Wirtz (2001), Rappa (2002).

[59] Choi and Stahl (1997), Porter (2001), Wirtz (2001).

nowadays is that they no longer need such large sums of capital to enter and compete within the market. This allows for both a lot more companies to participate and for those companies to compete on equal footing as the internet minimizes market imperfection.

The internet also facilitates lower distribution costs, more significant market share, and higher revenues.[60]

According to Hatton (2004), given the pressures from both consumers to lower prices and investors to maximize returns, intermediaries are being squeezed out. Between the reduction in transaction costs and the elimination of commissions, disintermediation appears to be a welcome and viable option.[61]

This illustrates the interaction between companies and the marketplace that they operate within. As consumers seek greater value for money, companies will make strategic decisions to increase profits.

Nowadays, it is common for customers to purchase goods and services from companies that have e-commerce websites online.[62]

An important debate within the industry and academia is whether or not any industries benefit particularly from the internet. Regarding disintermediation, Poon (2001) refers to the situation now being experienced by both suppliers and consumers as "new independence." The internet provides a single, sustainable electronic infrastructure that allows consumers and suppliers to exchange information and conduct business transactions.[63] While there are theorists who posit that the utilization of the benefits presented by the internet may lead to a kind of *reintermediation*, part of the aim of this book is to figure out why travel firms, in particular, refuse to adopt these beneficial innovations.

The term "reintermediation" refers to a school of thought that claims that projections of the complete removal of intermediaries may be premature.[64] According to this particular school of thought, a new kind of intermediary may develop. It is posited that this new intermediary, an "infomediary," will be responsible for brokering relationships between

[60] Laws (2001).

[61] Buhalis and Licata (2002).

[62] Ghosh (1998), Zappala and Gray (2006).

[63] Law et al. (2004).

[64] Palmer and McCole (1999).

producers and consumers online. The infomediary does this by learning about the consumer and the supplier and merging their interests online.

Agents will still be fulfilling their functions as counselors, but they now require an online presence to fulfill that function through. As a counsellor, they must become familiar with large amounts of information to assist the consumer in whittling down their choices and finding the optimal choices according to the consumer's preference.[65] This requires the ability to access information, assess quality, and provide expert advice. Users can access information, but they are usually overwhelmed by it. Infomediaries simplify the buying process for consumers by providing a platform to exchange information and make recommendations using that information, connecting the right buyers with the right sellers.

The debate between the proponents of disintermediation and the advocates for reintermediation is still ongoing, but this has led to a broadening of the discussion on technology adoption. From this growth have emerged concepts like technology access and questions such as how socioeconomic and cultural factors affect people's attitudes toward technology. One of the most interesting aspects of this widened debate, especially within post-internet discourse, is that of the global digital divide. Modern discourse now looks at newly relevant issues of information gaps, information inequality and information poverty across, and within, societies.

The Digital Divide

The Digital Divide refers to the situation whereby access to technology provides varying degrees of benefits and disadvantages across individuals. Current research aims to determine how the digital divide impacts technology adoption in small owner-managed entities. There have been, so far, four distinct approaches in analyzing the digital divide[66]:

- The Technocratic Approach
- The Social Structure Approach
- The Information Structure and Exclusion Approach
- The Modernization and Capitalism Approach.

[65] Lang (2000).

[66] Sassi (2005).

The Technocratic Approach assumes that the internet is an essential tool for modern life and commerce and that it will be a crucial instrument in overcoming social inequality. As such, they believe that it is incumbent upon the government to ensure access to the internet for all citizens. The Social Structure Approach emphasizes that, just as there is inequality in societies, there is also uneven access to the internet across societies, and each issue feeds into the other.[67]

The Information Structure and Exclusion Approach looks at how social segregation and marginalization due to poverty act as a barrier, preventing some groups from adopting new technologies and innovations as readily as other, more affluent groups. The Modernisation and Capitalism Approach argues that capitalism drives modernization and, hence, technology adoption. Castells (2000) posits that what has emerged is a new form of capitalism he refers to as "informational capitalism."

What these different approaches share in common is the acknowledgment and investigation of the inequality in ICT diffusion. Fundamentally, the only thing that separates them is their deliberation on causes and solutions.

This book mostly applies the Information Structure and Exclusion Approach and the Modernisation and Capitalism Approach. The leading proponents of each approach, Lash (1994) and Castells (2000) respectively, are concerned about how the global economy influences the digital divide in various countries, which in turn begs the question of how does the digital divide affect technology adoption in Caribbean countries specifically.

Even though the most fruitful period of the debate on the digital divide was between 1986 and 2005, there was little discussion on how it affected the tourism industry. Later, in 2010, Minghetti and Buhalis claimed that "tourists and destinations within developed countries and between developed and developing countries suffer from a multiplicity of technological divides (motivational, physical, informational, etc.), which lead to different levels of digital exclusion."[68]

It is clear that the digital divide also exists within the tourism and travel industries. In fact, given that these industries require interaction between

[67] Winston (1986), Sparks (2000), May (2002).

[68] Minghetti and Buhalis (2010, p. 278).

individuals and firms from different countries, it has been argued that the digital divide has more pronounced effects within these industries.

While academics and industry leaders have debated the digital divide since the 1970s, the evolution of the discussion came during the turn of the century with the introduction of the concept of "informational capitalism." Castells (2000) presents the argument that inequality and polarization within society are outcomes of informational capitalism, and that these issues should be resolved through public policy. Norris (2000) postulates that the digital divide is likely to widen even further given the challenges presented by global poverty. This is consistent with the existing literature. Spink and Cole (2001) surveyed the literature and noted that in 1975 over 700 papers had raised the issues of information inequality and information poverty.

This issue persists to this day, exacerbated by unequal access to information technologies.

The digital divide has been conceptualized in 3 ways:

- The Access Divide
- The Learning Divide
- The Content Divide.

The Access Divide was the initial conceptualization, but the last two are currently more accepted. James (2004) posits that access to technology is less of an issue now, relative to decades past, hence a narrowing of the access divide. Rogers (2003), on the other hand, states that it is the learning and content divides that are posing a particular challenge for some nowadays.

This perspective highlights the issue of matching content to audience needs. Innovators and designers of technology innovations are naturally going to create content and provide solutions that appeal to their audiences first. After some initial success, new material may be developed for external audiences. Still, there is already a lag in adoption as the learning curve for these external audiences is higher, given that the original content did not fit with their situations nor environments.

The focus on the content divide and the learning divide has revivified the discussions of contemporary theorists by looking at the surrounding political, socioeconomic, and cultural elements.[69] Now what is needed is a deeper analysis of these elements to see how they may be useful in bridging these divides.

So far, the discussions have revolved mainly around issues of access and public policy. Access to technology is critical, but it is not the sole issue. Even with access, some individuals may not adopt new technology because they simply are not comfortable with it. That being the case, what government bodies can do is to facilitate training in these technologies in order to change individual attitudes to technology. Given that they cannot influence the content, this may be the only viable option.

Recent additions to the discussion have been the emergence of *new media*[70] and *mobile technology*.[71] These new technologies have spread rapidly among developing nations and helped reduced the disparity between nations.

While inequality exists between nations, the focus of that inequality has shifted in recent years. While earlier research focused on the ethical perspective, later research was more concerned about the political and economic dimensions of inequality.[72] A significant development in the literature has been the shift from a discussion of fairness to one of practicality. Contemporary theorists note that information inequality has become a principal political issue in modern society.[73]

Not all theorists share the same perspective on the digital divide and how best to address it, so there are outliers.[74] Gunkel (2003), for instance, notes that critiques of the digital divide neither identify difficulties nor provide solutions but simply seek to determine what caused it. While those theorists may have a point, they also fail to realize that by analyzing the current situation's underlying preconditions, it is possible to better understand what is going on and determine how to resolve it.

The lion's share of research into the digital divide has been focused on the divide within societies, but now, perhaps influenced by the presence

[69] Wilson et al. (2003), Gyamfi (2005), Willis and Tranter (2006), Stump et al. (2008).

[70] Mansell (2002).

[71] Stump et al. (2008).

[72] Yu (2006).

[73] Feather (1998), Castells (2000), Norris (2000), Golding and Murdock (2001).

[74] Gunkel (2003), James (2004).

of the internet, the focus has shifted to technological inequality between countries.

The general argument that has been put forward is that those developed countries that control ICT development automatically have a distinct advantage.[75] James (2004) however, counter the trend by arguing that even though individual access may be limited in smaller developing countries, there is still innovation in the application of the technology itself. He states that this may be due partly to the influx of foreign direct investment by multinational corporations and partly by culture. Culture will also be explored as a factor of the digital divide and as an influence on technology adoption, especially how it relates to issues of leadership and resources.

Culture has been an object of study since the 1800s. It has been defined, generally, as a shared system of meaning,[76] where ideas that relate to values, norms, and traditions are exchanged, agreed upon and influence the group. Culture may be conceptualized at various levels—national, regional, local, organizational, consumer market-related, etc. In this book, we'll be focused on national and organizational culture and how these cultures influence the individual.

According to Fischer (2009), the shared meaning system of a cultural group influences its members through values, opinions, beliefs, and behavioral tendencies. As such, individuals are expected to act in accordance with their cultural system, and deviations from those norms carry negative consequences.[77] This naturally leads to discussions of cross-cultural differences.

Craig (2003) notes the significant point that while different behavioral tendencies are being studied, the same criteria being tested are the same criteria being used to select the studies' participants. For example, a study looking at a particular demographic assumes that there is general uniformity within that demographic.

Heinich (2010) counters this point by stating that the researcher does not impose cultural differences, but that systems of meaning are

[75] Mosaic Group (1998), Castells (2000), Norris (2000), Rogers (2003), Kirkman et al. (2002), Drori and Jang (2003).

[76] Fischer (2009).

[77] Toomela (2003).

what drives the research and allow the identification of differences across cultures and cultural groups.

Hofstede (1980) has been looking at the differences in work-related values between different countries, specifically along cultural dimensions. His model focuses on differences along four specific dimensions:

- *Power distance*: those who are powerless accept that, within their group, the distribution of power is unequal.
- *Masculinity v femininity*: while masculine systems tend to value money and success, feminine systems tend to be focused on nurturing and quality of life.
- *Individualism v collectivism*: as opposed to teamwork, individualism emphasizes each person's responsibility in taking care of himself and his family only.
- *Uncertainty avoidance*: the extent to which individuals are comfortable with risk.

Each of these is assessed when observing the interplay between owner-managers and their staff (power distance), result orientation (masculine), or staff empowerment (feminine), and risk aversion (uncertainty). The overall relationship orientation of society (individualism/collectivism) is also observed.

This model has been so effective that it has been applied to fields beyond that of its origination. Studies in management have benefited in particular, as Hofstede's discourse on cultural differences has been applied in analyzing organizational behavior across different national/cultural contexts. Similar models have also emerged since then.[78] While some have argued that there has been an over-reliance on the model,[79] it continues to be applied in academic work and researchers still utilize it as a template for understanding cultural differences. That is, until the turn of the century.

In 2002, McSweeny argued that Hofstede's model is fundamentally flawed. McSweeney questions whether culture can systematically cause differences in behavior between people from different countries. McSweeney's critique is challenged by Williamson (2002), who stated

[78] Schwartz (1992), Trompenaars (1993).

[79] Bhimani (2006), Harrison and McKinnon (1999), Redding (1994).

that Hofstede's research, grounded in the functionalist paradigm, was methodologically sound and it was McSweeney's paradigm that required clarity.

It has become an accepted truth that any work that assumes uniformity among the individuals within a society is flawed as there are observable differences among people within societies and these differences manifest in different behaviors.

This is, of course, not to claim that it is impossible to understand aspects of human activity or that it might be explained by the culture within which they live.

McSweeney's critique of Hofstede's four dimensions that they are too simplistic to explain national differences is ironic, given that it acknowledges that these differences are still useful in providing a basis for analysis. Contemporary authors continue to explore the impact of national culture on organizations and personal identity. They also acknowledge the relationships between organizational behavior and aspects of national culture. Specifically, with regard to understanding:

- multinational corporation and sales patterns[80]
- creativity and innovation[81]
- business practices[82]
- tourism[83]
- leadership[84]
- corporate strategy[85]
- business communication.[86]

The work of Westwood and Low (2003) is of particular interest to this study as it identifies a relationship between national culture and innovation, especially within the domain of creativity.

[80] Murphy (1999).

[81] Westwood and Low (2003).

[82] Tchaïcha and Davis (2005).

[83] Reisinger and Turner (2002).

[84] Elenkov and Manev (2005).

[85] Ulijn et al. (2000).

[86] Ibid.

It is surprising that, to date, the literature on culture and innovation focuses heavily on cultures within organizations[87] while neglecting the national culture that envelops it. As such, the relationship between organizational innovation adoption and national culture has either been ignored or, arguably, indirectly addressed through discourse on the global digital divide.[88]

One of the fundamental tenets of this book is leadership as a complementary driver of technology adoption. Naturally, this will also require an assessment of the influence of national culture on the concept of leadership. Elenkov and Manev (2005) posit that leadership factors and cultural context significantly impact top-management innovation influence. This claim is made on the basis that firms operate in a top-down paradigm and so rely on leadership from top executives who make the innovation adoption decisions. This coincides with the normative view within the literature that while leaders influence strategic decisions within firms,[89] it is the norms, values, and practices within a country's culture that influence innovation and its rate of diffusion[90] within the nation.

Ownership and Leadership

Organisational Theorists explore the relationship between the ownership of firms, decision-making within firms and the performance of those firms. Siebens (2002) defined ownership as having control of and obtaining the majority of the gains (shareholdings) from a firm's activities.

Decades of research into this area reflect the value placed in understanding this relationship,[91] but most of this research has been focused on ownership structures and their impact on performance. Less attention has been paid to the actual impact of leadership itself upon the behavior

[87] Sarros et al. (2008), Jaskyte and Kisieliene (2006), McLean (2005), Subramaniam and Ashkanasy (2001).

[88] Wilson et al. (2003), Gyamfi (2005), Willis and Tranter (2006), Stump et al. (2008).

[89] Hitt et al. (2001), Victorino et al. (2006), Ghobadian and O'Regan (2006), Falk (2008).

[90] Gibson and McDaniel (2010).

[91] Siebens (2002), Demsetz and Lehn (1985), Choa (1998), Cole and Mehran (1998), Villalonga and Amit (2006), Welch (2003).

of their firms.[92] This is ironic, given that it may be said that such behavior is a reflection of their owners.

Within the Organisational Theory literature, there are many divergent views regarding the impact of firm ownership upon firm performance. Past studies may be categorized by what they focus on, such as:

- Entrepreneurship
- Firm behavior
- Strategy
- Family owned businesses.

Porter (1990), for example, posits that the ownership of said firm influences a firm's goals. Daily and Dollinger more precisely state that a firm's ownership influences its size, strategy, and process. These observations have been quite useful in the debate as, in most cases, it is the owners that make the decisions regarding key strategic issues, and it is these decisions that ultimately determine whether a company succeeds or fails.

Several factors influence entrepreneurial behaviors themselves:

- Education
- Age
- Gender
- Number of family members.

Management behaviors have traditionally been regarded as being influenced primarily by management skills,[93] but a considerable body of research in management behavior has focused on the role of psychological factors.[94] Factors such as:

- Motivation
- Personality
- Goals (or Intention to grow).

[92] Ghobadian and O'Regan (2006).

[93] Wasilczuk (2000), Singh et al. (2001), DeMartino et al. (2006).

[94] Hamlin and Sawyer (2007), Frese et al. (2000).

As stated earlier, there has been considerable research into the ownership and leadership, specifically, of family owned businesses. With special attention being paid to how these businesses operate and the issues and challenges they face within particular industries.

Kowalewski et al. (2010) posit that firms with family CEOs outperform those with nonfamily CEOs. This, they state, is due to the CEO's relatively higher level of social capital and their greater personal investment in the business.

This framing is consistent with the work of Villalonga and Amit (2006). On the contrary, however, Tsao (2009) posits that family ownership may negatively affect performance due to, ironically, too much personal involvement or incompetence of family leaders. This shifts the focus of the discussion to the owner-managers of these firms.

Apart from being the majority shareholders of their respective businesses, owner-managers also engage in their business' daily operations.[95] Owner-managed firms tend to be relatively small companies,[96] while also being the owner-manager's sole mode of income. Such a combination of circumstances usually results in less bureaucracy when it comes to making decisions, but also engenders greater risk aversion, both of which influence the owner's managerial style. Managerial style generally refers to a manager's preferences in decision-making.

Whereas strategy focuses on a formalized method of reaching goals, style is concerned with a general, informal pattern of decisions in discretional situations.[97] It's related to decisions that embody:

- Risk-taking
- Change and innovation
- The aggressiveness of the approach used.

Carland et al. (2007) classified small business managers into two distinct types:

- Entrepreneurs
- Small business managers.

[95] Shailer (1994).

[96] Gallo and Christensen (2011).

[97] Covin and Slevin (1988).

When it comes to building strategies and preferences in behavior, entrepreneurs exhibit certain characteristics relative to managers,[98] such as:

- Higher levels of self-sufficiency
- Readiness for change
- Interest in innovation
- Competitive aggression
- Desire for achievement.

Meanwhile, general small business managers tend to exhibit characteristics such as[99]:

- Risk aversion
- Lack of innovation
- Passive behavior
- Reactive behavior.

Entrepreneurs will creatively utilize their resources to pursue profit and growth. At the same time, general small business managers will operate their businesses either as an extension of their personality, or pursue personal goals.[100] Management activities that implement the preferences of a manager in the operation of their business are termed as *tactical* or *operational behaviour*. These tactical behaviors are affected by the preferences of a decision-maker at the strategic level.[101]

A few examples of elements of practical business management activities would be[102]:

- Collecting information and learning new management skills
- Applying new technology or developing new products
- Quality control of products and service

[98] Carland et al. (2007), Covin and Slevin (1988).

[99] Covin and Slevin (1988).

[100] Carland et al. (2007).

[101] Slevin and Covin (1995).

[102] Page and Getz (1997), Dahles and Bras (1999), Morrison et al. (1999).

- Interaction with customers
- The treatment of people in business networks
- Cooperating with other organizations
- Employing family labor
- Long-term and papered financial planning
- Risk-taking.

Some authors, in an attempt to address firm ownership and performance, have found that firm ownership is a positive determinant of firm performance. This pronouncement addresses certain gaps in the previously existing literature. It naturally follows that if ownership is closely linked to performance, then an owner's influence on or prevention of strategy falls within these parameters.

George et al. (2005) extend the discussion, stating that the owners also influence firm internationalization. They argue that the ownership of small to medium-sized enterprises (SMEs) influences the proclivity to take risks. This is driven by the desire to increase both the scale of the company and the scope of internationalization efforts. This is particularly important as competitiveness warrants business exploration into new territories and markets.[103]

Recent research into firm ownership has attempted to expand the discourse from solely focusing on public corporations to also including private corporations and partnerships.[104] One particularly interesting finding of this expanded research is that private corporations and partnerships usually outperformed public corporations. This, they state, is due to the increased monitoring and span of control of the owners in these organizations. This contradicts the work of Durand and Vargas (2003), which suggests that this higher performance is due to greater organizational complexity.

What is consistent in organizational theory is that owners influence the key variables that affect firm performance. This influence is enhanced when they take on the role of owner-manager.[105] Kouzes and Posner (1987) define leadership as a reciprocal relationship that clearly separates

[103] Porter (1985), David (2007).

[104] Greenwood et al. (2007).

[105] Durand and Vargas (2003), Greenwood and Empson (2003).

those who influence from those who are influenced. While this definition provides a basic understanding of leadership, Hitt et al. (2001) describe the leader, in specific, as a catalyst for strategic change. Given this definition, it becomes critical to assess the dominant leadership categories.

In his seminal work, Burns (1978) introduced the concepts of transformational and transactional leadership. Several researchers have since attempted to expand on these concepts and, in particular, how they are measured.

In this research, leadership is emphasized as the key driver of technology adoption in small, owner-managed travel firms. As such, an evaluation of the two dominant classifications of leadership is vitally important. Unfortunately, research that focuses on leadership in the travel and technology industries is lacking.

Transformational leaders uplift the morale, motivation, and morals of their followers, while transactional leaders tend to cater to their followers' immediate self-interests. Transformational leadership also creates change and provides a framework for exploring attributes of change leadership.

According to Burns (1978), there is also a considerable distinction between the two forms of leadership. He articulates his position by stating that changes in the marketplace and workforce over the years have resulted in the need for leaders to become more transformational and less transactional to remain effective. Transformational leaders empower their followers by developing them into high-involvement individuals and teams focused on the following attributes:

- Quality
- Service
- Cost-effectiveness
- Quantity of output.

While this research does not directly address the question of transformational leadership vs transactional leadership, it does identity new initiatives in the matter of leadership, such as the adoption of innovations. For example, the creation of high-involvement individuals and cost-effectiveness are required for successful technology adoption.

It must be stated that there have been many issues surrounding transformational/leadership literature. Pawar (2003), for example, argues that there are concerns regarding how transformational leadership has been conceptualized. One such issue relates to the relationship between the two types of leadership. There are essentially three schools of thought on this relationship.

Burns (1978) viewed both forms of leadership as being at opposite ends of the leadership continuum. Bass (1985) identified certain weaknesses in a second conceptualization of the paradigm. He acknowledged that while both forms of leadership were distinct, there may be varying relationships between them. The third school of thought regarding this relationship[106] states that transformational leadership is merely an extension of transactional leadership.

The primary reason for this is that some fundamental activities require leaders to entice followers to conform to procedures, while others require significant transformational skills. For this work, these two forms of leadership will be treated as a hierarchical conceptualization, with transactional leadership providing the foundation and necessary preconditions for transformational leadership.

A second issue refers to attempts to separate the concept of transformational leadership from charismatic leadership. Burns (1978) originally viewed charismatic leadership as a type of transformational leadership. Subsequent research has since identified that charisma is a central element of transformational leadership.

Transformational leadership has been described as moving the follower beyond immediate self-interests through "idealized influence, inspiration, intellectual stimulation or individualised consideration." The element that relates to charisma, coined by Bass, is the idealized influence that was used to frivolously described popular individuals. This was very useful, as "idealised influence" occurs "when the leader envisions a desirable future, articulates how it can be reached, sets an example to be followed, sets high standards of performance, and shows determination and confidence. Followers want to identify with such leadership."[107]

While all of the four constructs above will be examined, *intellectual stimulation* is the most applicable construct in technology adoption. This is due to the fact that studies in innovation and adoption focus on creativity and new ideas. This construct is displayed when the leader helps his followers to become more innovative and creative. This is

[106] Hollander (1993).

[107] Ibid., p. 11.

operationalized and measured by a leader's openness to their followers' ideas.

The remaining constructs all address the relationship and kinship that the followers have with their leaders. While these all result in the followers aligning themselves to the goals of the leader, only ***intellectual stimulation*** addresses cognitive and behavioral changes in the followers. This occurs via the leader's constant encouragement to generate new ideas. Hence, a leader who practices intellectual stimulation creates an atmosphere for the generation of new ideas at every level of the firm. The resulting atmosphere stimulates creative thoughts and behaviors among employees within the firm, leading to the acceptance and understanding of innovative changes within that organization.

Intellectual stimulation is a requirement in changing people's perspectives on things such as work and the process of innovation. This research attempts to fill the gap in the existing literature on the issue of the interplay between leaders and the general context of the organization.

It must be acknowledged that different types of organizations require varying degrees of transformational leadership.[108] This highlights the need for research that clarifies whether or not certain changes that are imposed by leaders are warranted in their respective organizational contexts. A second issue is whether or not the given organizational context supports or opposes transformational leadership behavior.[109] The focus of this book, owner-managers who determine and implement policies unilaterally, is not so dependent upon the second point. This is because owner-managers within these firms can create a bespoke organizational structure that caters to their own style of management, suggesting that these leaders have greater influence over the organizational context than vice versa.

Leaders play a crucial role in determining which innovations to introduce.[110] This is significant as innovations give a firm an edge over its competitors,[111] or assist in providing more efficient operations. Leaders are also responsible for resource handling, and their willingness to provide

[108] Pawar (2003).

[109] Pettigrew (1987), Pawar and Eastman (1997).

[110] Victorino et al. (2006).

[111] Porter (1985), David (2007).

resources directly impacts quality and performance.[112] This fits well into the overall discussion of the resource-based view. It also provides an added dimension, suggesting that a firm with a broad resource base is not inherently competitive because it is still subject to decisions and directions of leadership. Competitiveness may be determined by a firm's resources and the effective use of those resources by the firm's leadership.

The Resource-Based View

The Resource-based analysis method has existed since the early 1950s[113] but became more popular three decades later when Wernerfelt (1984) explored the usefulness of assessing organizations from the resource side rather than the product side, which was the formerly dominant approach.[114]

The resource perspective focuses on the assets, tangible, and intangible, that a firm owns. Caves (1980) defines a firm's resources as those assets that are semi-permanently tied to the firm. These may include[115]:

- Machinery
- Capital
- Brand names
- In-house knowledge of technology
- Employment of skilled personnel
- Trade contracts
- Efficient procedures.

This definition marked the beginning of a detailed assessment of a firm's internal capabilities and, by extension, its ability to compete. There has been much discussion about the competitive edge that is provided by the difference in resources between firms. Barney (2014) articulates that these differences may be long-lasting, providing a sustained advantage

[112] Lewis and McCann (2004).

[113] Penrose (1959).

[114] Porter (1980).

[115] Wernerfelt (1984).

for some firms. In fact, other perspectives on resource-based assessment, whether neoclassical economics,[116] evolutionary economics,[117] or structure-conduct-performance (SCP) based theories of competitive advantage,[118] also share this opinion.

Neoclassical theorists purport that greater resources and capabilities, referred by them as factors of production, enable companies to provide greater supply. They also purport that this supply is elastic, suggesting that a rise in demand will result in an increase in the price and supply of the resource. Due to the evolution of resources, some resources may be considered inelastic due to the length of time it would take to develop them.[119] An example of this would be human resources. This inelasticity may, however, be a source of sustained competitive advantage as other firms may take some time to develop their own resources.[120]

Aligning this resource-based view with Nelson and Winter's (1982) work on evolutionary economics would suggest that different firms conduct business in different ways. Ineffective and inefficient operations, as measured by differences in input and output, decrease a firm's competitiveness. The implication is that those firms with the necessary financial, human and technical resources to drive their operations are likely to have an advantage.

The most popular set of discussions surrounding resource-based perspectives, however, are more closely positioned to the SCP-based models of competitive advantage.[121] These models more directly advance the notion that a firm's attributes and resources determine its performance and sustained advantage.

According to resource-based theorists, the more resources a firm holds relative to its competitors, barring any irrational actions, the more competitive it may be.[122] The idea of acting rationally points to the fact

[116] Ricardo (1817).

[117] Nelson and Winter (1982).

[118] Porter (1980).

[119] Barney (2001).

[120] Peteraf (1993).

[121] Barney (2001).

[122] Penrose (1959), Colli (2011), Teece (1982), Wernerfelt (1984), Peteraf (1993), Mowery et al. (1998).

that the actions of the current resource holders may affect the costs and revenues of the future holder of those resources.

The resource position barrier is similar to Porter's (1995) entry barriers, but there is one key difference. Market incumbents implement entry barriers to block new market entrants, while resource position barriers are more focused on providing barrier situations to those that are already in the market. The resource position barrier, therefore, provides a basis upon which firms that are already competing with each other may outdo one another[123] by using their resource position to cement their lead. What is evident in these discussions is that it is common for a firm's resources to dictate what can and cannot be done.

Peteraf (1993) more closely aligns resources with firm performance as the resource-based model concerns itself with the internal accumulation of assets and asset specificity. This presupposes that firms have different internal capabilities and resources, a fundamental assumption when assessing a firm's competencies and resources for the purpose of planning strategies.

As no two companies have the same resources, different companies require different strategies. Firms with broader resource bases are therefore more likely to pursue diversification,[124] enabling them to decide, for example, upon which platforms to capitalize. Diversification may refer to:

- Diversification of products
- Diversification of markets
- Diversification of processes.

While firm-specific capabilities have gained remarkable attention, management scholars have extended the resource-based view to look at more than just how firms compete but also how they collaborate.[125]

There has also been an exploration of how a firm's resources influence business partner choice. Mowery et al. (1998) also looked at how the technological capabilities of a firm evolved relative to its assets and firm-specific capabilities and posited how the technological overlap plays

[123] Kraajenbrink et al. (2010).

[124] Montgomery and Hariharan (1991), Teece (1982), Wernerfelt (1984), Williamson (1985), Mahoney and Pandian (1992).

[125] Teece (1982), Mowery et al. (1998).

a role in partner selection. More broadly applicable, however, is the fact that firms with significant resource bases are particularly attractive for collaboration. The extant literature does, unfortunately, fail to expound on which categories of resources are seen as most attractive, much less the consideration of variation in what is considered attractive across different industries.

Essentially, resource-based discussions currently focus on a firm's competencies, capabilities, and strategy. They argue that a firm's competencies are a result of the assets that a firm has at any single point in time, hence, the ability to prolong or sustain any relative advantage will depend on the type of resources and type of industry. While the resource-based view is not suitable for providing managerial prescriptions,[126] it does point out why some firms have a distinct advantage over others, a feature that can be of great use to managers.[127]

Brouthers et al. (2008), in extending the discussion to an international context, suggest that differences in the institutional environments between nations may affect its applicability. They argue that when the differences in institutional environments are applied to the resource-based view, this allows it to be more robust and stronger in its application. As such, context specificity should play a role when assessing firm performance and strategy considerations.

Emergent Frameworks

While specific gaps in existing research have been highlighted, the general gap in terms of approach points to the fact that approach has only been explored in terms of simplistic relationships between a limited set of variables, i.e., the singular relationships between personal factors, firm strategy, and resources. As a result, very little is known about the complex relationship between ownership/leadership and technology adoption, as the discussion so far has neglected the complex interconnectedness of a multiplicity of variables.

The key argument of this research is that firm technology adoption is affected by multiple factors working concurrently. These factors are

[126] Barney (2001).

[127] Kraajenbrink et al. (2010).

highlighted through a distilled framework that draws attention to some antecedents, such as:

- Ownership/Leadership
- Firm strategy
- Resources
- Culture
- The Digital Divide.

This framework, taken within the context of the Theory of Organisational Decision-Making, identifies the key driver as being Ownership/Leadership. This work, therefore, explores:

a. how these personal factors are influenced by macro issues such as culture and the global digital divide
b. how they, in turn, influence firm strategy and resources
c. And how these factors, ultimately, affect technology adoption decisions within firms.

Iterations of the conceptual frameworks that have emerged from an analysis of the gaps in the literature are illustrated below in Figs. 2.1 and 2.2.

Figure 2.1 identifies fragmented relationships that exist within the literature. The researcher has, however, brought these together to demonstrate that whereas one factor may directly influence another, it may indirectly affect other factors within the technology adoption discourse.

Due to the fragmentation present in previous works, there have been some gaps in addressing technology adoption in firms. Figure 2.2 illustrates a more coherent framework that places technology adoption decisions within the broader theory of decision-making, while also highlighting relationships that may be explored in future research.

This research further emphasizes the leadership role as owner-managers of small travel firms have a great deal of autonomy in decision-making. This second-order iteration is created as an input framework to be tested, based on the literature and the gaps in the first-order iteration.

The conceptual framework in Fig. 2.2 weaves together the various concepts discussed in the literature review. It establishes that the path to technology adoption within firms involves multiple factors. Those factors

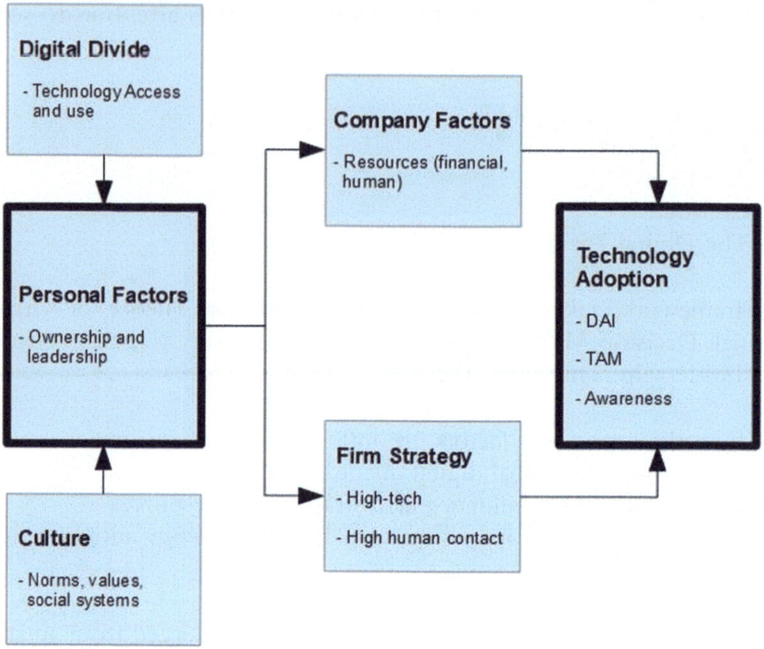

Fig. 2.1 Firm technology adoption framework: first-order iteration (*Source* Author's creation)

that are external to the firm, e.g., culture and the digital divide, provide the context that the firm must operate within. This influences how the ownership of firms evolve and, even more so, how leadership attitudes are formed through socialization and education.

The types of entrepreneurs and leaders that emerge will directly influence how firms are structured and the strategic choices that are made and supported by resource allocation. These factors drive the behavior of the firm, including those behaviors relevant to innovation. This is a product of organizational culture, which is, in turn, developed through strategy and policy decisions.

The research on firm behavior identifies a clear relationship between how a firm behaves and its performance, as measured by its profitability and competitiveness. The literature points to systematic approaches to the assessment of firm performance and is geared toward informing strategy

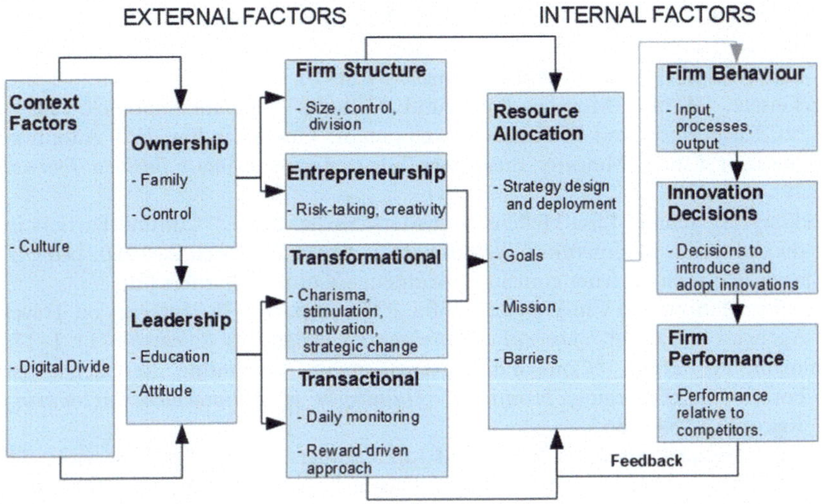

Fig. 2.2 Technology decision-making input framework (*Source* Author's creation)

formulation. This will either support previous technology decisions or result in changes to previous decisions, as demonstrated in the feedback loop illustrated in the input framework (Fig. 2.2).

References

Afuah, A., and C.L. Tucci. 2003. *Internet Business Models and Strategies: Text and Cases* (Vol. 2). New York: McGraw-Hill.

Bagozzi, R.P. 2007. "The Legacy of the Technology Acceptance Model and a Proposal for a Paradigm Shift." *Journal of the Association for Information Systems* 8(4): 3.

Barnett, Martin, and Craig Standing. 2001. "Repositioning Travel Agencies on the Internet." *Journal of Vacation Marketing* 7: 143.

Barney, J.B. 2001. "Resource-Based Theories of Competitive Advantage: A Ten Year Retrospective on the Resource-Based View." *Journal of Management* 27: 643.

———. 2014. *Gaining and Sustaining Competitive Advantage*. London: Pearson Higher Education.

Bass, Bernard. M., Bruce J. Avolio, Dong I. Jung, and Yair Berson. 2003. "Predicting Unit Performance by Assessing Transformational and Transactional Leadership." *Journal of Applied Psychology* 88(2): 207–218.

Beckinsale, Martin, Monder Ram, and Nicholas Theodorakopoulos. 2011. "ICT Adoption and Ebusiness Development: Understanding ICT Adoption amongst Ethnic Minority Businesses." *International Small Business Journal* 29: 193–219.

Beekhuyzen, Jenine, Liisa Hellens, and Mark Siedle. 2005. "Cultural Barriers in the Adoption of Emerging Technologies." Accessed March 2, 2010. http://www.ucd.smartinternet.com.au/Documents/Cultural_Barriers.pdf.

Bennett, Marion, and Chi-Wen Lai. 2005. "The Impact of the Internet on Travel Agencies in Taiwan." *Journal of Tourism and Hospitality Research* 6(1): 1–17.

Bhimani, A., 2006. "Comparative Management Accounting Research: Past Forays and Emerging Frontiers." *Handbooks of Management Accounting Research* 1: 343–363.

Bigne, Enrique J., Joaquín Aldas, and Luisa Anderu. 2008. "B2B Services: IT Adoption in Travel Agency Supply Chains." *Journal of Services Marketing* 22(6): 454–464.

Brouthers, Keith D., Lance E. Brouthers, and Steve Werner. 2008. "Resource-Based Advantages in the International Context." *Journal of Management* 34(2): 189.

Buhalis, Dimitrios, and Maria Cristina Licata. 2002. "The Future eTourism Intermediaries." *Journal of Tourism Management* 23(3): 207–220.

Buhalis, Dimitrios, and Ourania Deimezi. 2004. "E-Tourism Developments in Greece: Information Communication Technologies Adoption for Strategic Management of the Greek Tourism Industry." *Tourism and Hospitality Research* 5(2): 1467–3584.

Burns, James M. 1978. *Leadership*. New York: Harper and Row.

Carland, J.W., F. Hoy, W.R. Boulton, and J.A.C. Carland. 2007. "Differentiating Entrepreneurs from Small Business Owners: A Conceptualization." In *Entrepreneurship*, 73–81. Berlin and Heidelberg: Springer.

Castells, Manuel. 2000. *End of Millennium*. 2nd ed. Oxford: Blackwell.

Caves, Richard E. 1980. "Industrial Organization, Corporate Strategy and Structure." *Journal of Economic Literature* 18(1): 64–92.

Chandler, Alfred. 2003. *Strategy and Structure: Chapters in the History of the American Industrial Enterprise*. Fairless Hills: Beard Books.

Choa, Myeong-Hyeon. 1998. "Ownership Structure, Investment and the Corpo rate Value: An Empirical Analysis." *Journal of Financial Economics* 47(1): 103–121.

Choi, Soon-Yong, and Dale Stahl. 1997. *The Economics of Electronic Commerce*. Indiana: Macmillan Technical Publishing.

Chorn, Norman. 1994. "The New Logistics Organization." In *The Gower handbook of Logistics and Distribution Management*. 4th ed., edited by John Gattorna, 501–508. Aldershot, England: Gower Publishing.

Christopher, Martin. 1986. *The Strategy of Distribution Management*. Oxford: Heinemann Professional Publishing.

———. 1992. *Logistics and Supply Chain Management: Strategies for Reducing Cost and Improving Service*. London: Prentice Hall.

———. 1998. *Logistics and Supply Chain Management: Strategies for Reducing Cost and Improving Service*. London: Prentice Hall.

Cole, Rebel A., and Hamid Mehran. 1998. "The Effect of Changes in Ownership Structure on Performance: Evidence from the Thrift Industry." *Journal of Financial Economics* 50(3): 291–317.

Colli, A. 2011. "Business History in Family Business Studies: From Neglect to Cooperation? *Journal of Family Business Management* 1: 12.

Covin, Jeffery G., and Dennis Slevin. 1988. "The Influence of Organization Structure on the Utility of an Entrepreneurial Top Management Style." *Journal of Management Studies* 25(3): 217–234.

Cox, Andrew. 1998. "Relational Competence and Strategic Procurement Management: Towards an Entrepreneurial and Contractual Theory of the Firm." *European Journal of Purchasing and Supply Management* 2: 57–70.

Craig, Anita P. 2003. "Culture and the Individual." *Theory and Psychology* 13(5): 629–650.

Crotts, John, Abdul Aziz, and Andrew Raschid. 1998. "Antecedents of Supplier's Commitment to Wholesale Buyers in the International Travel Trade." *Tourism Management* 19(2): 127–134.

Cyert, Richard M., and James March. 1963. *A Behavioural Theory of the Firm*. Englewood Cliffs, NJ: Prentice Hall.

Dahles, Heidi, and Karin Bras. 1999. "Entrepreneurs in Romance: Tourism in Indonesia." *Annals of Tourism Research* 26: 267–293.

Dale, Crispin. 2003. "The Competitive Networks of Tourism E-mediaries: New strategies, New Advantages." *Journal of Vacation Marketing* 9: 109.

DeMartino, Richard, Robert Barbato, and Paul Jacques. 2006. "Exploring the Career/Achievement and Personal Life Orientation Differences between Entrepreneurs and Non Entrepreneurs: The Impact of Sex and Dependents." *Journal of Small Business Management* 44(3): 350–369.

Daniel, Elizabeth, Hugh Wilson, and Andrew Myers. 2002. "Adoption of E-Commerce by SMEs in the UK: Towards a Stage Model." *International Small Business Journal* 20(3): 253–270.

Davis, Fred D. 1989. "Perceived Ease of Use, and User Acceptance of Information Technology." *MIS Quarterly* 13(3): 319–339.

David, Fred D. 2007. *Strategic Management: Concepts and Cases*. Upper Saddle River, NJ: Prentice Hall.

DeHayes, Daniel W., and Robert L. Taylor. 1972. "Making Logistics Work in a Firm." *Business Horizons* 24(1): 1–26.

Demsetz, Harold, and Kenneth Lehn. 1985. "The Structure of Corporate Ownership." *Journal of Political Economy* 93(6): 1155–1177.

Drori, Gili, and Yong Suk Jang. 2003. "The Global Digital Divide: A Sociological Assessment of Trends and Causes." *Social Science Computer Review* 21: 144–161.

Dulle, Frankwell W., and M.K. Minishi-Majanja. 2011. "The Suitability of the Unified Theory of Acceptance and Use of Technology (UTAUT) Model in Open Access Adoption Studies." *Information Development* 27(1): 32–45.

Durand, Rodolphe, and Vicente Vargas. 2003. "Ownership, Organization, and Private Firms' Efficient Use of Resources." *Strategic Management Journal* 24(7): 667–75.

Dyerson, Romano, and G. Harindranath. 2007. "ICT Adoption and Use by SMEs in the UK: A Survey of South East." Paper Presented at the Portland International Center for Management of Engineering and Technology (PICMET), Portland, Oregon, USA.

Eames, Donald, and Gregory Norkus. 1988. "Developing Your Procurement Strategy." *Cornell Hotel and Restaurant Administration Quarterly* 29: 30–33.

Elenkov, Detelin, and Ivan Manev. 2005. "Top Management Leadership and Influence on Innovation: The Role of Socio-cultural Context." *Journal of Management* 31(3): 381–402.

Falk, Martin. 2008. "Effects of Foreign Ownership on Innovation Activities: Empirical Evidence for Twelve European Countries." *National Institute Economic Review* 204: 85–97.

Feather, John. 1998. *The Information Society: A Study of Continuity and Change*. 2nd ed. London: Library Association Publishing.

Fischer, Ronald. 2009. "Where Is Culture in Cross Cultural Research? An Outline of a Multilevel Research Process for Measuring Culture as a Shared Meaning System." *International Journal of Cross Cultural Management* 9: 25–49.

Fishbein, Martin, and Icek Azjen. 1975. *Belief, Attitude, Intention and Behaviour: An Introduction to Theory and Research*. Reading, MA: Addison-Wesley.

Frese, Michael, Marco van Gelderen, and Michael Ombach. 2000. "How to Plan as a Small Scale Business Owner: Psychological Process Characteristics of Action Strategies and Success" *Journal of Small Business Management* 38: 1–18.

Fuchs, Matthias, Wolfram Hopken, Andreas Foger, and Martin Kunz. 2009. "E-Business Readiness, Intensity and Impact: An Austrian Destination Management Organization Study." *Journal of Travel Research* 49(2): 165–178.

Gallo, Peter J., and Lisa Jones Christensen. 2011. "Firm Size Matters: An Empirical Investigation of Organizational Size and Ownership on Sustainability-Related Behaviors." *Business Society* 50: 315–349.

Gentner, D., B. Stelzer, and L. Brecht. 2017, June. "Analyzing Customers' Readiness for Digital B2B Business Models." In *ISPIM Innovation Symposium*, 1. The International Society for Professional Innovation Management (ISPIM).

Gattorna, John. 1994. *The Gower Handbook of Logistics and Distribution Management*. 4th ed. Aldershot, England: Gower Publishing.

George, Gerard, Johan Wiklund, and Shaker Zahra. 2005. "Ownership and the Internationalization of Small Firms." *Journal of Management* 31(2): 210–233.

Ghobadian, Abby, and Nicholas O'Regan. 2006. "The Impact of Ownership on Small Firm Behaviour and Performance." *International Small Business Journal* 24(6): 555–584.

Ghosh, Shikhar. 1998. "Making Business Sense of the Internet." *Harvard Business Review* 124–135.

Gibson, Cristina, and Dana McDaniel. 2010. "Moving Beyond Conventional Wisdom Advancements in Cross-Cultural Theories of Leadership, Conflict, and Teams." *Perspectives on Psychological Science* 5: 450–462.

Golding, Peter, and Graham Murdock. 2001. "Digital Divide: Communications Policy and Its Contradictions." *New Economy* 8(2): 110–115.

Grandon, Elizabeth, and J. Michael Pearson. 2004. "Electronic Commerce Adoption: An Empirical Study of Small and Medium US Businesses." *Information and Management* 42: 197–216.

Gray, Colin. 2006. "Stage Models of ICT Adoption in Small Firms." In *Impact of E-commerce on Consumers and Small Firms*, edited by Salvatore Zappala and Colin Gray, 3–20. Aldershot: Ashgate.

Greenwood, Royston, and Laura Empson. 2003. "The Professional Partnership: Relic or Exemplary Form of Governance?" *Organization Studies* 24(6): 909–933.

Greenwood, Royston, David Deephouse, and Stan Li. 2007. "Ownership and Performance of Professional Service Firms." *Organization Studies* 28(2): 219–238.

Grieger, M. 2003. "Electronic Marketplaces: A Literature Review and a Call for Supply Chain Management Research." *European Journal of Operational Research* 144(2): 280–294.

Gunkel, David. 2003. "Second Thoughts: Toward a Critique of the Digital Divide." *New Media and Society* 5(4): 499–522.

Gyamfi, Alexander. 2005. "Closing the Digital Divide in Sub-Saharan Africa: Meeting the Challenges of the Information Age." *Information Development* 21(1): 22–30.

Hamel, Gary, and C. K. Prahalad. 1994. *Competing for the Future*. Boston: Harvard Business School Press.

Hamlin, Robert, and Jenny Sawyer. 2007. "Developing Effective Leadership Behaviours: The Value of Evidence-based Management." *Business Leadership Review* 4(4): 1–16.

Harland, Christine M. 1996. "Supply Chain Management, Purchasing and Supply Management, Logistics, Vertical Integration, Materials Management and Supply Chain Dynamics." In *Blackwell Encyclopedic Dictionary of Operations Management*, edited by Nigel Slack. Oxford, UK: Blackwell.

Harrison, Graeme, and Jill McKinnon. 1999. "Cross-Cultural Research in Management Control Systems Design: A Review of the Current State." *Accounting, Organizations and Society* 24: 483–506.

Hatton, Gerry. 1994. "Designing a Warehouse or Distribution Centre." In *The Gower Handbook of Logistics and Distribution Management*, edited by John Gattorna, 175–193. 4th ed. Aldershot, England: Gower Publishing.

Hatton, Mike. 2004. "Redefining the Relationships: The Future of Travel Agencies and the Global Agency Contract in a Changing Distribution System." *Journal of Vacation Marketing* 10: 101.

Heinich, Nathalie. 2010. "What Does 'Sociology of Culture' Mean? Notes on a Few Trans-Cultural Misunderstandings." *Cultural Sociology* 4: 257–265.

Heung, Vincent C. S. 2003. "Barriers to Implementing E-Commerce in the Travel Industry: A Practical Perspective." *International Journal of Hospitality Management* 22(1): 111–118.

Hitt, Michael A., R. Duane Ireland, and Robert E. Hoskisson. 2001. *Strategic Management: Competitiveness and Globalization*. 4th ed. Mason, OH: South Western Publishing.

Hoffman, Eric, and Paul Roman. 1984. "Information Diffusion in the Implementation of Innovation Process." *Communication Research* 11(1): 117–140.

Hofstede, Geert. 1980. *Culture's Consequences: International Differences in Work-related Values*. Beverly Hills, CA: Sage.

Hollander, Edwin P. 1993. "Legitimacy, Power and Influence: A Perspective on Relational Features of Leadership." In *Leadership Theory and Research: Perspectives and Directions*, edited by Martin M. Chemers, and Roya Ayman, 29–48. San Diego: Academic Press.

Van den Hooff, Bart, Jasper Groot, and Sander de Jonge. 2005. "Situational Influences on the Use of Communication Technologies: A Meta-Analysis and Exploratory Study." *Journal of Business Communication (1973)* 4(2): 4–27.

Houghton, Kathryn, and Heidi Winklehofer. 2004. "The Effect of Website and E-commerce Adoption on the Relationship between SMEs and Their Export Intermediaries." *International Small Business Journal* 22: 369.

James, Jeffrey. 2004. "The Global Digital Divide in the Internet: Developed Countries Constructs and Third World Realities." *Journal of Information Science* 31(2): 114–123.

Jaskyte, Kristina, and Kisieliene Audrone. 2006. "Organizational Innovation: Comparison of Nonprofit Human-Service Organizations in Lithuania and the United States." *International Social Work* 49(2): 165–176.

Karamustafa, Kurtulus. 2000. "Marketing-Channel Relationships: Turkey's Resort Purveyors' Interactions with International Tour Operators." *Cornell Hotel and Restaurant Administration Quarterly* 41(4): 21–32.

Keller, Kenneth. 2008. "From Here to There in Information Technology: The Complexities of Innovation." *American Behavioral Scientist* 51(1): 97–106.

Kirkman, Geoffrey, Peter Cornelius, Jeffrey Sachs, and Klaus Schwab. 2002. *Global Information Technology Report 2001–2002*. Oxford: Oxford University Press.

Kotler, Philip, and Gary Armstrong. 2008. *Principles of Marketing*, 12th ed. Upper Saddle River, NJ: Pearson-Prentice Hall.

Kouzes, James M., and Barry Posner. 1987. *The Leadership Challenge: How to Keep Getting Extraordinary Things Done in Organizations*. San Francisco, CA: Jossey Bass.

Kowalewski, Oskar, Oleksandr Talavera, and Ivan Stetsyuk. 2010. "Influence of Family Involvement in Management and Ownership on Firm Performance: Evidence from Poland." *Family Business Review* 23(1): 45–59.

Kraajenbrink, Jeroen, J. C. Spender, and Aard Groen. 2010. "The Resource-Based View: A Review and Assessment of Its Critiques." *Journal of Management* 36: 349.

Kuan, Kevin K. Y., and Patrick Chau. 2001. "A Perception-Based Model for EDI Adoption in Small Businesses Using Technology-Organization-Environment Framework." *Information and Management* 38: 507–521.

Kuhnert, Karl, and Philip Lewis. 1987. "Transactional and Transformational Leadership: A Constructive/Developmental Analysis." *The Academy of Management Review* 12(4): 648–657.

Kwon, Winston, Ian Clarke, and Ruth Wodak. 2009. "Organizational Decision-Making, Discourse, and Power: Integrating across Contexts and Scales." *Discourse and Communication* 3: 273–302.

Lang, Tania. 2000. "The Effect of the Internet on Travel Consumer Purchasing Behaviour and Implications for Travel Agencies." *Journal of Vacation Marketing* 6: 368.

Langley, Ann, Henry Mintzberg, Patricia Pitcher, Elizabeth Posada, and Jan Saint-Macary. 1995. "Opening Up Decision Making: The View from the Black Stool." *Organisation Science* 6(3): 260–279.

Lash, Scott. 1994. "Reflexivity and Its Doubles: Structure, Aesthetics, Community." In *Reflexive Modernization*, edited by Ulrich Beck, Anthony Giddens, and Scott Lash, 110–173. Cambridge: Polity Press.

Latzer, Michael. 2009. "Information and Communication Technology Innovations: Radical and Disruptive?" *New Media Society* 11(4): 599–619.

Law, Rob, Kenith Leung, and James Wong. 2004. "The Impact of the Internet on Travel Agencies." *International Journal of Contemporary Hospitality Management* 16(2): 100–107.

Laws, Eric. 2001. "Distribution Channel Analysis for Leisure Travel." In *Tourism Distribution Channels: Practices, Issues and Transformations*, edited by Dimitrios Buhalis, and Eric Laws, 53–72. London: Continuum.

Lederer, A. L., Maupin, D. J., Sena, M. P., & Zhuang, Y. 1998, June. "The Role of Ease of Use, Usefulness and Attitude in the Prediction of World Wide Web Usage." In *Proceedings of the 1998 ACM SIGCPR Conference on Computer Personnel Research*, 195–204.

Legris, Paul, John Ingham, and Pierre Collerette. 2003. "Why Do People Use Information Technology? A Critical Review of the Technology Acceptance Model." *Information and Management* 40: 191–204.

Lewis, Barbara, and Pamela McCann. 2004. "Service Failure and Recovery: Evidence from the Hotel Industry." *International Journal of Contemporary Hospitality Management* 16(1): 6–17.

Lippert, Susan K., and Miles Davis. 2006. "A Conceptual Model Integrating Trust into Planned Change Activities to Enhance Technology Adoption Behaviour." *Journal of Information Science* 32(5): 434–448.

Loudon, David L., and Albert J. Della Bitta. 1993. *Consumer Behaviour: Concepts and Applications*. 4th ed. New York: McGraw-Hill.

Lumsdon, Les M., and Jonathan S. Swift. 1999. "The Role of Tour Operators in South America: Argentina, Chile, Paraguay and Uraguay." *International Journal of Tourism Research* 4(1): 429–439.

Mahoney, Joseph T., and J. Rajendran Pandian. 1992. "The Resource-Based View Within the Conversation of Strategic Management." *Strategic Management Journal* 13: 363–380.

Mansell, Robin. 2002. "From Digital Divides to Digital Entitlements in Knowledge Societies." *Current Sociology* 50(3): 407–426.

Marler, Janet H., Xiaoya Liang, and James Hamilton Dulebhon. 2006. "Training and Effective Employee Information Technology Use." *Journal of Management* 32(5): 721–743.

May, Christopher. 2002. *Information Society: A Sceptical View*. Cambridge: Polity Press.

McLean, Laird D. 2005. "Organizational Culture's Influence on Creativity and Innovation: A Review of the Literature and Implications for Human Resource Development." *Advances in Developing Human Resources* 7(2): 226–246.

Minghetti, Valeria, and Dimitrios Buhalis. 2010. "Digital Divide in Tourism." *Journal of Travel Research* 49(3): 267–281.

Mochrie, Robert, Laura Galloway, and David Deakins. 2006. "The Value of Internet Forums to Small Rural Businesses." In *Impact of e-Commerce on Consumers and Small Firms*, edited by Salvatore Zappala and Colin Gray, 81–100. Aldershot: Ashgate.

Moital, Miguel, Roger Vaughan, and Jonathan Edwards. 2009. "Using Involvement for Segmenting the Adoption of E-commerce in Travel." *Service Industries Journal* 29(5): 723–739.

Montgomery, Cynthia A., and Sam Harihan. 1991. "Diversified Expansion by Large Established Firms." *Journal of Economic Behaviour* 52(1): 71–89.

Morrison, Alison, Michael Rimmington, and Clare Williams. 1999. *Entrepreneurship in the Hospitality, Tourism and Leisure Industries*. Oxford: Butterworth-Heinemann.

Mosaic Group. 1998. "The Global Diffusion of the Internet Project: An Initial Inductive Study." Accessed January 16, 2010. www.agsd.com/gdi97/gdi97.html.

Mowery, David C., Joanne E. Oxley, and Brian Silverman. 1998. "Technological Overlap and Interfirm Cooperation: Implications for the Resource-Based View of the Firm." *Research Policy* 27: 507–523.

Murphy, William H. 1999. "Hofstede's National Culture as a Guide for Sales Practices across Countries: The Case of a MNC's Sales Practices in Australia and New Zealand." *Australian Journal of Management* 24(1): 37–58.

Nelson, Richard R., and Sidney Winter. 1982. *An Evolutionary Theory of Economic Behaviour and Capabilities*. Cambridge: Harvard University Press, 195–307.

Norris, Pipps. 2000. "Digital Divide? Civic Engagement, Information Poverty and the Internet in Democratic Societies." Accessed November 9, 2010.

Page, Stephen, and Don Getz. 1997. *The Business of Rural Tourism: International perspectives*. London: Thomson Business Press.

Palmer, Adrian, and Patrick McCole. 1999. "The Virtual Re-intermediation of Travel Services: A Conceptual Framework and Empirical Investigation." *Journal of Vacation Marketing* 6(1): 33–47.

Patricia, Dodu. 2008. "The Internet, Threat or Tool for Travel Agencies?" Accessed April 9, 2010. http://steconomice.uoradea.ro/anale/volume/2008/v2-economy-andbusiness-administration/017.pdf.

Pawar, Badrinarayan S. 2003. "Central Conceptual Issues in Transformational Leadership Research." *Leadership and Organization Development Journal* 24(7): 397–406.

Pawar, Badrinarayan S., and Kenneth Eastman. 1997. "The Nature and Implications of Contextual Influences on Transformational Leadership: A Conceptual Examination." *Academy Management Review* 22: 80–109.

Pearce, Douglas G. 2009. "Beyond Tiers: A Network Approach to Tourism Distribution." *Tourism Analysis* 13: 517–530.

Penrose, Edith T. 1959. *The Theory of the Growth of the Firm*. New York: Wiley.

Persson, Goran. 1978. "Organisation Design Strategies for Business Logistics." *International Journal of Physical Distribution and Materials Management* 8(6): 287–297.

Peteraf, Margaret A. 1993. "The Cornerstones of Competitive Advantage: A Resource-Based View." *Strategic Management Journal* 14: 179–191.

Pettigrew, Andrew M. 1987. "Context and Action in the Transformation of Firms." *Journal of Management Studies* 24: 649–670.

Poon, Auliana. 2001. "The Future of Travel Agents." *Travel & Tourism Analyst* 3: 57–80.

Poon, Simpson, and Mathew Joseph. 2001. "A Preliminary Study of Product Nature and Electronic Commerce." *Marketing Intelligence & Planning* 19(7): 493–499.

Porter, Michael. 1980. *Competitive Strategy*. New York: The Free Press.

———. 1985. *Competitive Advantage*. New York: The Free Press.

———. 1990. *The Competitive Advantage of Nations*. London: Macmillan.

———. 1995. *Wettbewerbsstrategien (Competitive Strategy)*. New York: Campus.

———. 2001. "Bewährte Strategien Werden mit dem Internet Noch Wirksamer." *Harvard Business Manager* 5: 64–81.

Porter, Michael, and Victor Millar. 1985. "How Information Gives You Competitive Advantage." *Harvard Business Review* 63(4): 149–160.

Presutti, William. 2003. "Supply Chain Management and E-procurement: Creating Value-Added in the Supply Chain." *Industrial Marketing Management* 32: 219–226.

Rappa, Michael. 2002. "Business Models on the Web." Accessed February 7, 2010. http://digitalenterprise.org/models/models.html.

Rayport, Jeffrey F. 1995. "Exploiting the Virtual Value Chain." *Harvard Business Review* 73(6): 75–85.

Redding, S.Gordon. 1994. "Comparative Management Theory: Jungle, Zoo or Fossil Bed?" *Organization Studies* 15: 323–359.

Reisinger, Yvette, and Lindsay W. Turner. 2002. "Cultural Differences between Asian Tourist Markets and Australian Hosts: Part 2." *Journal of Travel Research* 40: 385–395.

Ricardo, David. 1817. *Principles of Political Economy and Taxation*. London: J. Murray.

Ricketts, Martin. 1994. *The Economies of Business Enterprise*. Hempstead: Harvester Wheatsheaf.

Reve, Torger. 1990. "The Firm as a Nexus of Internal and External Contracts." In *The Firm as a Nexus of Treaties*, edited by Masahiko Aioki, 136–188. London: Sage.

Rogers, Everett. 1962. *Diffusion of Innovations*. New York: The Free Press.
———. 1976. "New Product Adoption and Diffusion." *Journal of Consumer Research* 2: 290.
———. 1983. *Diffusion of Innovations*. New York: The Free Press.
———. 1994. *A History of Communication Study—A Biographical Approach*. New York: The Free Press.
———. 2003. *Diffusion of Innovations*. 5th ed. New York: The Free Press.
Rushton, Alan, and John Oxley. 1993. *Handbook of Logistics and Distribution Management*. London: Kogan Page Limited.
Ryals, Lynette, and Andre Humphries. 2007. "Managing Key Business-to-Business Relationships: What Marketing Can Learn from Supply Chain Management." *Journal of Service Research* 9: 312.
Ryan, Bryce, and Neal Gross. 1943. "The Diffusion of Hybrid Seed Corn in Two Iowa Communities." *Rural Sociology* 8: 15–24.
Saffu, Kojo, and John Walker. 2008. "Strategic Value and Electronic Commerce Adoption among Small and Medium-sized Enterprises in a Transactional Economy." *Journal of Business and Industrial Marketing* 23(6): 395–404.
Sarros, James C., Brain K. Cooper, and Joseph Santora. 2008. "Building a Climate for Innovation through Transformational Leadership and Organizational Culture." *Journal of Leadership & Organizational Studies* 15: 145–158.
Sassi, Sinikka. 2005. "Cultural Differentiation or Social Segregation? Four Approaches to the Digital Divide." *New Media and Society* 7(5): 684–700.
Schary, Philip, and Tage Larsen. 1998. *Managing the Global Supply Chain*. Copenhagen: Copenhagen Business School Press.
Schwartz, Shalom. 1992. "Universals in the Content and Structure of Values: Theoretical Advances and Empirical Tests in 20 Countries." *Advances in Experimental Social Psychology* 25: 1–65.
Shailer, Gregory. 1994. "Capitalists and Entrepreneurs in Owner-Managed Firms." *International Small Business Journal* 12: 33–41.
Siebens, H. 2002. "Concepts and Working Instruments for Corporate Governance." *Journal of Business Ethics* 39(1–2): 109–116.
Simon Herbert Alexander. 1957. *Models of Man*. New York: Wiley.
Singh, Surendra P., Ruthie G. Reynolds, and Safdar Muhammad. 2001. "A Gender-Based Analysis of Micro and Small Enterprises in Java, Indonesia." *Journal of Small Business Management* 39(2): 174–182.
Slevin, D.P., & J.G. Covin. (1995). "Entrepreneurship as Firm Behavior: A Research Model." *Advances in Entrepreneurship, Firm Emergence, and Growth* 2: 175–224.
Sparks, Colin. 2000. "The Distribution of Online Resources and the Democratic Potential of the Internet." In *Cultural, Economic and Policy Foundations for*

Media Openness and Diversity in East and West, edited by Jan van Cuilenburg, and Richard van der Wurff, 229–254. Amsterdam: Het Spinhuis.

Spink, A., and C. Cole. 2001. "Information and Poverty: Information-Seeking Channels Used by African American Low-Income Households." *Library & Information Science Research* 23(1): 45–65.

Stansfield, Mark, and Kevin Grant. 2003. "An Investigation into Issues Influencing the Use of the Internet and Electronic Commerce among Small-Medium Sized Enterprises." *Journal of Electronic Commerce Research* 4(1): 15–33.

Straub, Evan. 2009. "Understanding Technology Adoption: Theory and Future Directions for Informal Learning." *Review of Educational Research* 79: 625–649.

Stump, Rodney L., Wen Gong, and Zhan Li. 2008. "Exploring the Digital Divide in Mobile-Phone Adoption Levels across Countries: Do Population Socioeconomic Traits Operate in the Same Manner as Their Individual-level Demographic Counterparts?" *Journal of Macromarketing* 28: 397–412.

Subramaniam, Nava, and Neal Ashkanasy. 2001. "The Effect of Organisational Culture Perceptions on the Relationship between Budgetary Participation and Managerial Job-Related Outcomes." *Australian Journal of Management* 26(1): 35–54.

Tatum, B. Charles, Richard Eberlin, Carin Kottraba, and Travis Bradberry. 2003. "Leadership, Decision-Making and Organizational Justice." *Management Decision* 41(10): 1006–1016.

Tchaïcha, Jane D., and Mark M. Davis. 2005. "The Impact of Culture on Technology and Business: An Interdisciplinary, Experiential Course Paradigm." *Journal of Management Education* 29(5): 738–757.

Teece, David J. 1982. "Toward an Economic Theory of the Multiproduct Firm." *Journal of Economic Behaviour and Organization* 3: 39–63.

Teo, Thompson, Sijie Lin, and Kee-hung Lai. 2009. "Adopters and Non-adopters of E-procurement in Singapore: An Empirical Study." *Omega* 37(5): 972–987.

Toomela, Aaro. 2003. "How Should Culture Be Studied?" *Culture & Psychology* 9(1): 35–45.

Tsao, Chiung-Wen. 2009. "Founding-Family Ownership and Firm Performance: The Role of High-Performance Work Systems." *Family Business Review* 22(4): 319–332.

Trompenaars, Fons. 1993. *Riding the Waves of Culture: Understanding Cultural Diversity in Business*. London: Economist Books.

Ulijn, Jan, Dan O'Hair, Mathieu Weggeman, Gerald Ledlow, and Thomas Hall. 2000. "Innovation, Corporate Strategy, and Cultural Context: What Is the Mission for International Business Communication?" *Journal of Business Communication* 37(3): 293–317.

Victorino, Liana, Rohit Verma, Gerhard Plaschka, and Chekitan Dev. 2006. "Service Innovation and Customer Choices in the Hospitality Industry." *Managing Service Quality* 15(6): 555–576.

Villalonga, B., and R. Amit. 2006. How Do Family Ownership, Control and Management Affect Firm Value? *Journal of Financial Economics* 80(2): 385–417.

Virolainen, Veli. 1998. "A Survey of Procurement Management Development in Industrial Companies." *International Journal of Productions Economics* 56: 677–688.

Vishwanath, Arun, and Gerald Goldhaber. 2003. "An Examination of the Factors Contributing to Adoption Decisions among Late-Diffused Technology Products." *New Media and Society* 5(4): 547–572.

Vrana, Vasiliki, and Costas Zafiropoulos. 2006. "Tourism Agents' Attitudes on Internet Adoption: An Analysis from Greece." *International Journal of Contemporary Hospitality Management* 18(7): 601–608.

Warden, Stuart, and S. Tunzelana. 2004. "E-Commerce: A Critical Review of SMME Organizational Barriers in Tourism." Paper presented at the Proceedings of 6th WWW Applications Conference 2004, Rand Afrikaans University, Johannesburg, South Africa.

Wasilczuk, Julita. 2000. "Advantageous Competence of Owner/Managers to Grow the Firm in Poland: Empirical Evidence." *Journal of Small Business Management* 38(2): 88–94.

Welch, Emma. 2003. "The Relationship between Ownership Structure and Performance in Listed Australian Companies." *Australian Journal of Management* 28(3): 287–306.

Wernerfelt, Birger. 1984. "A Resource-Based View of the Firm." *Strategic Management Journal* 5: 171–180.

Werthner, Hannes, and Stefan Klein. 1999. *Information Technology and Tourism: A Challenging Relationship*. Vienna: Springer Verlag.

Westwood, Robert, and David Low. 2003. "The Multicultural Muse Culture, Creativity and Innovation." *International Journal of Cross Cultural Management* 3(2): 235–259.

Williamson, Dermot. 2002. "Forward from a Critique of Hofstede's Model of National Culture." *Human Relations* 55(11): 1373–1395.

Williamson, Oliver E. 1979. "Transaction Cost-Economies: The Governance of Contractual Relations." *Journal of Law and Economics* 22: 232–261.

———. 1985. *The Economic Institutions of Capitalism*. New York: Free Press.

———. 1990. "The Firm as a Nexus of Treaties: An Introduction." In *The Firm as a Nexus of Treaties*, edited by Masahiko Aioki, 136–188. London: Sage.

Williamson, Oliver E., and Scott E. Masten. 1999. *The Economics of Transaction Costs*. Cheltenham: Edward Elgar Publishing.

Willis, Suzanne, and Bruce Tranter. 2006. "Beyond the 'Digital Divide': Internet Diffusion and Inequality in Australia." *Journal of Sociology* 42(1): 43–59.

Wilson, Kenneth R., Jennifer S. Wallin, and Christa Reiser. 2003. "Social Stratification and the Digital Divide." *Social Science Computer Review* 21: 133–143.

Winston, Brian. 1986. *Misunderstanding Media*. London: Routledge and Kegan Paul.

Wirtz, Bernd W. 2001. *Electronic Business*. Wiesbaden: Gabler Verlag.

Wöber, Karl, and Ulrike Gretzel. 2000. "Tourism Managers' Adoption of Marketing Decision Support Systems." *Journal of Travel Research* 39: 172–181.

Woiceshyn, Jaana. 2000. "Technology Adoption: Organizational Learning in Oil Firms." *Organization Studies* 21(6): 1095–1118.

Yu, Liangzhi. 2006. "Understanding Information Inequality: Making Sense of the Literature of the Information and Digital Divides." *Journal of Librarianship and Information Science* 38(4): 229–252.

Zappala, Salvatore, and Colin Gray. 2006. *Impact of e-Commerce on Consumers and Small Firms*. Aldershot: Ashgate.

Zappala, Salvatore, and Guido Sarchielli. 2006. "Climate for Innovation, Attitude to Internet and ICT Adoption in Small Firms." In *Impact of e-Commerce on Consumers and Small Firms*, edited by Salvatore Zappala, and Colin Gray, 35–50. Aldershot: Ashgate.

Zhou, Yuqiong. 2008. "Voluntary Adopters Versus Forced Adopters: Integrating the Diffusion of Innovation Theory and the Technology Acceptance Model to Study Intra-organization Adoption." *New Media Society* 10: 475.

CHAPTER 3

Internal Firm Factors: An Examination of Travel Companies Resource Base

The tourism industry is a major contributor to Caribbean economies such as Jamaica, Trinidad and Tobago, and the Bahamas. The industry employs 32.8, 8.5, and 52.5% of the population, respectively.[1] Considering the significant relationship between the tourism industry and employment within these countries, the industry's level of development should be examined to understand its ability to adapt in a data driven and technological environment.

It is posited that information communication technology (ICT) is a significant contributing factor in enhancing the tourism product marketed by these island States.[2] Used appropriately, ICT can provide high-quality tourism products for tourists and administrators of tourism. These benefits can be seen in products which enables tourists to access tourism services and products from anywhere, at any time. Conversely, administrator's ability to reach target customers across the world by mobile phones and web technologies has increased.[3] Successful integration of ICT in areas such as customer reservation to office automation can potentially lead to customer retention and customer loyalty, which inevitably lead to growth and development of economies.

[1] The World Travel and Tourism Council (2019). Retrieved from https://wttc.org/Research/Economic-Impact.

[2] O'Connor (2008).

[3] Bethapudi (2013).

A. Spencer, *Technology Adoption in the Caribbean Tourism Industry*,
https://doi.org/10.1007/978-3-030-61584-0_3

This study is timely in light of ICT proliferation in the global landscape; therefore, consideration must be given to the tourism industry in these countries and consequently their travel firms given that they operate within the confines of the technology that is available in the societies. In so far as growth and development of economies, travel firms play an integral role; their ability to adopt and successfully integrate ICT in their process can inherently lead to increased quality of tourism products and services as well as increased customer availability, retention, and loyalty.

Deeper analysis was obtained using Jamaica as a case example; 31 of the 43 travel agencies within Jamaica. The firms interviewed subscribed to the following selection criteria—they consisted of owner-managers (those owners who engage in the daily operational activities of the firm). The selection of owner-manager firms was due to their autonomous decision-making nature. These firms however, varied in size and operational periods. More than half (67.7%) of the firm sampled operated for more than 15 years while the remaining 16.1% operated for less than 10 years.

In comparison travel companies in the Bahamas on average have been opened for 20 years. Like firms in Jamaica, research findings revealed that the firm had a long-standing history of operation within the country.

To determine the size of the firms, the number of employees were used in Jamaica 87.1% of the firms had fewer than 15 employees and 71% had less than 10 employees. Within the Bahamas firms had on average 6 employees. Based on the aforementioned features the firms may be categorized as mature, small firms.

The years of operation for these firms might signify why most of them are opposed to new innovations. Lynskey (2004) contended that a relationship exists between the number of years a firm has been in operation and the innovative activity that firm pursues. On one hand, some scholars argued that age and experience may lead to more competitive behaviors, while the opposing argument is that newer agile firms may be more open to cutting-edge approaches. The latter would seem to be more applicable in this research as owner-managers in older firms highlighted that their traditional methods continue to work therefore the prevailing perception is that the need for change is not evident. The following statement by a respondent conveys as much: "We have been in this business for almost 20 years and we have survived harder times than these. For the last 10 years we have used SABRE and it has worked well for us. I think that these challenges will soon pass just like they did when we faced commission cuts in 2001."

Another key finding was that, the client bases of these firms remained fairly constant, with little new business being generated. Additionally, respondents recognized that the client base was decreasing rather than increasing, as their clientele (primarily baby boomers), seemed to be booking less through them while Generation X and Y were utilizing alternative means of booking online. Notably, the markets served by all these firms were the outbound leisure and business markets but none of the firms served the inbound leisure or business markets; due to its lack of commission. The excessive emphasis on the outbound leisure market can be attributed to that these business owners served most of their clients for over 10 years and that they felt that there was a high level of loyalty. This expression by an owner-manager reflects the sentiments of all those interviewed. He expressed, "Most of my customers have been with me for a long time and now they seem to be booking less, which maybe because of the global economic crisis or just the fact that they are older than they were and are naturally traveling less. At the same time, we tend not to get a lot of young people booking with us because they probably do it themselves."

The interviews with owner-managers revealed a unanimous reliance among all the participants on GDS, specifically SABRE and AMADEUS. In this regard, the majority (96.8%) of firms perceived their technology use as very high. While most firms recognized the importance of technology for sales and marketing, they failed to see its importance outside of GDSs. In light of this stance, 77.4% of Jamaican firms thought company websites were unimportant and an even larger proportion (87%) viewed the online market to be very unimportant. Unsurprisingly, none of the firms had a social media account (Facebook, MySpace, or twitter), and the internet was primarily used for emailing.

Only one owner-managed company in Trinidad and Tobago has a website and felt that online market was important while the remaining companies perceive the online market to be unimportant.

Companies that created websites in the past noted at the time it seemed like a great idea, one that would positively impact the company's visibility. Subsequent to the website creation, firms recognized that they lack the personnel to maintain and update it. Conversely, an owner-manager stated "creating a website is one thing but maintaining it is completely different matter." In its initial stage the websites provided potential customers with information, much of which is now outdated. Of the firms with websites a portion were interested in visibility and providing general information about the company and its location while others sought to provide marketing information such as package deals. Nonetheless, in most instances' customers made bookings on the telephone or at the office. This infers that firms utilized technology (websites) to influence traditional offline booking practice. The adoption of technology in this minimalistic approach highlights the conservative, traditional, and short-sightedness of firms. This is brought home by one owner-manager who stated "when designing the website, we did not consider doing complete bookings online because that is complicated and risky. We just wanted to become more visible so that potential customers would see us and call us to make bookings."

A noteworthy finding however, was the fact that all of these firms agreed that the variables identified by Eason (1998) were applicable. More than 90% of respondents agreed that the internet promotes faster work, makes work simpler, takes less effort, and improves the quality of output. Yet, firms were in disagreement as to whether the internet was cost-efficient and improves their image among competitors. While the operational benefits were readily identifiable by firms, they were less convinced of its value from a strategy and resource perspective. To bolster this stance, all the respondents acknowledged an internet investment as medium (45.2%) or high risk (54.8%). This connotes their uncertainty about the return on investment regarding the internet. This therefore, gives rise to the three types of perceived risk: psychological risk (Eastlick and Lotz 1999), time risk (Tan 1999), and financial risk. All three were applicable as owner-managers were apparently worried about the risk of losing time and money but also about the psychological risk of becoming dependent on the internet. The perceived diminishing nature of the internet by owner-managers is echoed by a respondent in the following statement. "I have seen too many companies in other industries become so technological that they lose some of their customer service quality and I do not want my company to become so dependent that we forget who we are."

Most firms' strategic decisions were customer driven, in collaboration with variables such as speed, convenience, accuracy, and security[4] which were of importance to the majority (80.6%) of their customers. Significantly, personal interaction was the only variable which all the firms

[4] Wöber and Gretzel (2000).

agreed was important to their customers. This service as articulated by the respondents could not be provided by the internet. As such their lack of adoption of the internet beyond GDS can be perceived as a strategic decision in an effort to meet the demands of their clientele. As aforementioned firms associate the value of technology to its utility with the GDS, therefore it is no surprise that all the firm's front office operations use GDS to serve their markets. This is very significant in highlighting their attitude to investing in technology. One respondent stated that: "The airlines give us these systems for free as well as training in how to use them so we do not have to invest too much or face the risk of losing money when things do not work out." Further investigation indicated that all of the agencies use computers for back-office accounting and administrative functions through software called "pear tree." This automated system facilitates the efficient handling of payables and receivables.

Traditional firms' reluctance toward adopting new technology is due to the perceived and actual cost of familiarizing employees with these technologies. The barrier then seems to be one of learning to use the internet.[5] Issues of ease of use[6] can be addressed with greater internet training. However, owner-managers overwhelmingly ascribe little importance in executing this kind of training. This lack of implementation can be attributed to their preference to use systems where free training is provided. One research participant noted "there would be additional cost in training employees and I am not sure that we would increase revenue as a result. At least with the GDS training is provided at no additional cost to us so even if there are no significant returns then we are not seriously affected."

The aforementioned statement substantiates the notion that the degree to which new technologies are adopted by these firms, is to foster greater operational focus rather than a strategic decision. The respondents myopic outlook did not detect any benefit to increased training in the use of internet technologies and chose to facilitate technology training only if this is free. These firm's strategy and resources were addressed with image among competitors and monetary savings, these factors align with the literature reviewed which influence adoption of innovations. Image was the only factor in this section on relative advantage for which there was no

[5] Rogers (2003), James (2004), Stump et al. (2008).

[6] Davis (1989).

ambiguity. There was a high level of disagreement among sample, 80.6% disagreed or totally disagreed that such a benefit existed. This was due in part to their perception of local operators as their competitors and these companies were equally as indecisive about the utilization of the internet especially for front office operations. One respondent expressed, "I do not think that our local competitors are using these technologies either and so whether I use it or not they will not think I have a better image. In any case they do not know what I am using or not using and I don't pay a lot of attention to what they use."

The lack of technology use among competitors for sales and marketing highlights owner-managers minimalistic approach utilizing technology. Noticeably, non-utilization of technology tools by competitors for sales and marketing provides a strong argument as to why its use probably yields a competitive advantage. The most distinct of these benefits is that of monetary savings. Owing to the fact that these are small firms with limited resources, the management structure for each firm is fairly simple—the owner is the manager and provides supervision for a few employees. The returns on investment therefore, was directly related the owner-manager's livelihood. One owner-manager stated, "my company is my life investment and so anything that I spend on for the agency must guarantee that I either make more money or save some money. I am really not sure that I would save or make any money from increased internet use." Concerns surrounding the efficient use of resources and competitive advantage are interrelated to strategy and resource allocation debate which will emerge after the case study analysis which summarizes the characteristics of the highest and one of the lowest adopters.

Case Examples

Case A

Within this company internet usage is defined by regular emailing, online browsing, and minimal online sale but with the intention to grow this platform. Led by an avid internet user, the owner-manager utilizes the internet for personal online shopping and social media use; this leader also has three children who are avid users. He notes this as being influential in his decision to establish an online platform in the company.

He states:

> I buy things online all the time like gadgets and books and I have always felt that it is an excellent way to reach customers. I have delayed using it in the business for a while because I figured that my client base was not ready. I think the time is right at this point as the client base is becoming more savvy. In any case I can try to reach people outside the current client pool.

> I currently use Facebook to connect with friends and I am not using it in the business as yet because I need to find out more about whether customers view it as a serious business tool or just as something to do for fun. I am open to the idea though and will look into it.

The owner-manager's experience with technology impacts his willingness to adopt several forms of innovative technological tools in the business. It also fosters a high degree of openness to exploring options. This openness is positively influenced by an university education in business studies, which allowed for an understanding of various business models.

> Back in my days at the University of the West Indies I learned how to analyze market needs on a broader scale through needs assessments. This kind of business knowledge can be applied to any industry and is not limited to just travel.

> I also had the privilege to work in banking and finance as well as real estate, and both of these industries faced hard times so I understand that in difficult times, which as you know the travel agencies are facing, we need to find new strategies to survive.

Notably, experience garnered in other industries also influence the manager's approach and perspective, he noted it made him less risk averse and "even though I live in a society that is not very open to technological investments, something new must be tried in times of crisis." The most profound statement made by this leader was that "you can't expect to do the same thing over and over and expect different results." The characteristics of this owner-manager reflects that of a market leader who positions himself to be proactive rather than reactive while seeking out new opportunities thwart impending challenges.

Case B

This case represents low adopting firms and embodies features of firms in this category. Internet use is restricted to emailing and the primary tool for sales is the GDS which was acquired at no cost. Therefore, investment in incurring cost to invest in technology and the internet is very low. Owner-managers within this category has been in the industry for the last 25 years and perceive the technology revolution to have a diminishing effect on service quality in the industry. One manager reminisced on his experience in the travel industry prior to the availability of cutting-edge technology, stating "When I first started in this industry, I used to work for Air Jamaica and then I went to Martin's Travel. Afterwards I opened my own agency because I was impressed with the level of service of these companies. Back then we didn't have all this online craze and we believed in personalized service."

This manager felt that there was a great deal of incompatibility with technology and maintaining the personality of the business, meaning the human element. It is critical to note however, that her narrow focus may be influenced by her level of education as her entrance into the travel industry occurred immediately after high school with no accompanying formal education or training. Her experience therefore limits her willingness to invest in technology without evidence of the benefits of its use. Compounded by the fact that her child is an adult who does not reside at home, this owner-manager is risk averse to the internet and highlights this by expressing "I would be wasting my time to try to sell online to my clients because most of them do not like using the internet to do serious business; furthermore they like having that personal contact and they wouldn't trade that for anything."

The personal traits of these owner-managers at both ends of the spectrum play a critical role in informing strategy and decisions pertaining to resource allocation. These are important internal elements of the firm in technology adoption studies as previously highlighted in Fig. 2.2. Each of which will be discussed in light of the findings from the qualitative component of the research.

Firm Strategy

Porter (1985) body of work which analyses business strategy in a competitive environment[7] is frequently referenced among strategic management theorists. A recurring argument in this body of work is that strategy development must take into consideration competitors and consumers. This approach promises a well informed and formulated, dynamic and robust competitive strategy. While at the core of any strategy is competitiveness, for the objectives of this research strategy will be conceptualized in three ways to reflect the supply chain. These are overall competitive strategy, procurement strategy, and distribution strategy. The role that each of these has in relation to technology adoption will be examined.

Porter (1985, 2001) posits that a company formulates a strategy to gain competitive advantage, it achieves this by responding to the following forces:

The findings largely support the work of Porter in terms of strategy. According to Porter (1985, 2001) a company develops its business strategies to obtain a competitive advantage, it achieves this by responding to five primary forces:

- Power of buyers
- Threat of substitute products/services
- Power of suppliers
- The threat of new entrants
- Rivalry among existing firms within an industry.

When assessed in relation to the aforementioned forces, firms in the study faced no or little threat of new entrants to the market given the steady decline in the number of firms operating in the environment. New entrants to the market space like Expedia were not deemed as a threat nor perceived as a competitor by the firms. Instead, rivalry among existing firms within the local market was a very important factor. The primary strategy competitive advantage used by rivaling firms was that of undercutting of service fees, this led to significant levels of mistrust. This mistrust was expressed by an owner-manager who stated, "as a group we decided that service fees should be J$2100 and there are some

[7] See for example David (2007), Kim et al. (2004), Stonehouse and Snowdon (2007), Poon (1993), Wynne et al. (2001).

companies who have secretly been charging as little as J$1500. At this point we cannot even trust the association to regulate the industry." The association the owner-manager was referring to is the Jamaica Association of Travel Agencies (JATA) of which all the firms were members. An owner-manager shed some light on the dynamics of the current market environment when he stated, "the president of our travel association (JATA) is the manager of the largest travel agency which has been buying all of our small firms. To have someone like this as the president means that the association is no longer looking out for the best interest of smaller, under-privileged firms like mine." In light of the reduction in commission by the suppliers a collective decision was reached to provide service at a selected minimum charge. However, some firms decided to provide services below the agreed floor price. This inevitably led to mistrust and absence of longer-term strategies to gain competitive advantage.

Suppliers and buyers' power are quite evident in the study. Major suppliers—airlines and their actions drastically impacted the number of agencies in Jamaica, Trinidad and Tobago, and Bahamas with the number declining from 105 to 43; 11 to 5 and within a decade, respectively. The reduction of commission from 9 to 6% by American Airlines, British Airways, and Air Jamaica in 2000 led to Jamaican travel agencies foiled attempt to boycott the sale of American Airline tickets sale in response to the 3% reduction in commission. American Airline which accounts for roughly 70% of air traffic into the region, remained steadfast in their outlook that direct bookings were practical course of action. Air Jamaica and British Airways implemented similar cuts two months later which further thwarted travel agencies position. The power of buyers is best seen on the decision to continue operation with minimal technological innovation premised on the perception that buyers are not necessarily interested in high technology platforms.

A prevailing theme among respondents surrounded whether or not some forms of technology adoption were compatible with their sales and distribution strategy. They expressed that this was driven by consumer needs and preferences. Interestingly, the belief held by managers is that a high-tech strategy paired with a hi-touch one were incompatible, this view is incongruent with the postulations by theorists[8] that a multi-platform approach to distribution will yield sound business models. The owner-managers concerns were found to be more in line with perceived psychological risk of being too reliant on the internet and the fading essence of their customer interface,[9] "if we rely too much on the technology our business loses its personality," one manager stated. Therefore, it becomes apparent that assurance of online and offline distribution channels compatibility must precede increased internet adoption among these owner-managers.

A variable which was not found in the literature but surfaced from the interviews with owner-managers was accountability. Owner-managers expressed that their clientele prefer to have someone culpable when their plans or reservations went awry, "they like to have someone to quarrel with if their booking is messed up and they can't do that with the net, emphasized an owner-manager. The presence of someone who is culpable for errors then said person would be responsible to implement corrective measures. Contrastingly, with online transactions managers believed that their "low-tech market" suffered from transaction loneliness.

In the process of determining the channels to use to distribute services, these firms were operating with an information deficit. Firstly, none of the owner-managers surveyed knew their competitor's distribution methods, as a result their claim that customers were technologically inept were unfounded. Secondly, they ignored the transitioning needs of their current clientele as well as the novel needs of the emerging markets by being singularly focused on the current platform used by their customers.

The companies under investigation for the study are intermediaries and while examining their strategies for distribution are important, analysis of their approaches to buying from suppliers is critical. The concept of strategic purchasing addresses relationships formed in business to business transactions. Eames and Norkus (1988) urge buyers to formulate a strategy comprised of these elements: self-analysis, vendor analysis, vendor-performance analysis, and program review. Yet it argued that the supply chain has undergone a quiet revolution unknown to business marketers, where focus has shifted to fostering longer-term relationships with few key suppliers rather than the traditional approach which

[8] See for example Buhalis and Licata (2002), Hatton (2004).

[9] Eastlick and Lotz (1999).

placed emphasis on cost-effective transactions.[10] This view is in contrast to several authors[11] previously mentioned and highlights a shift in the commonly held perception that the key driver is transaction cost.

There was no clear procurement strategy for the companies sampled; this lack of strategy can be attributed to a small supplier's pool in relation to the kind of air travel in demand (namely from Jamaica to the United States) and limited suppliers' diversification in areas such as tours, cruises, and hotels. Owner-managers firm diversification into services like cruises and accommodations would facilitate their renewed level of importance through a greater influence within the inbound travel. Considering that cruises and hotels offers five times more commission than airlines in the Caribbean. It may be emphatically stated that while a clear procurement strategy has an effect on these firms who utilize GDS systems made available by suppliers; the development of innovative procurement strategy is not facilitated by the current situation. The firm strategy debate regarding technology adoption foe owner-managed companies must be examined in areas of competitive and distribution strategies. However, these have surfaces as effects of broader factors like leadership in small owner-managed firms.

Leadership and Strategy Formulation

The characteristics which define the firms under study are that they were small, owner-managed companies in an information intensive market. These factors are extremely influential in the level of autonomy owners possess over their daily operations. Their action tends to meet the best interest of the firm, family, and personal interests.

The small size of owner-managed firms connotes a simple organizational structure regarding reporting relationships. As a consequence, owner-managers possess greater autonomous capacity; most of the firms examined also had no accountability to a board of directors. Typically, decision-making is unilateral and reflects the preferences of these owner-managers. This is essential in an information intensive industry where firms utilize minimal cutting-edge technology. One owner-manager stated that decisions regarding which cutting-edge technology to adopt

[10] Ryals and Humphries (2007).

[11] Ricketts (1994), Williamson (1979, 1990), Reve (1990).

were made as follows "If I see the benefit of some new innovation then I will check to see if I have the money to introduce it. If I am convinced that it is something that I absolutely need then I slowly introduce by telling employees about it and then doing orientation about it." Another expressed, "Well I make the decisions since I am the one who will be answerable to my family if the business fails. I sometimes ask for the advice of friends who may have some experience with a tool or with a similar business but at the end of the day it comes down to my decision."

Strategy formulation in these firms fall solely in the purview of owner-managers and reflects their experiences. As aforementioned in Chapter 7 the level of ICT adoption is directly related to the manager's personal technology involvement. Therefore firms with technologically savvy leaders had a greater tendency to incorporate technologies within their distribution strategies. The distinguishing feature of the most technologically advanced firm was a developed online selling platform, however its use was being revitalized while researching the utility of social media as a marketing tool. The manager of this company stated, "the power of the internet cannot be ignored. Markets are changing. As a consumer I use the internet to purchase books, music and travel products and so I know that in the same way that this works for companies that I buy from, it can work for my business too."

Conversely, owner-managers whose highest form of technology was the GDS system were inexperienced with ICT and maintained traditional approaches. One leader expressed his reluctance to introducing new technological tools stating "I have to make decisions that will enhance my firm and I am not comfortable introducing things that I am not familiar with. If I use the technology and I am comfortable with how it works then I might use it for the business. I don't use most of these new technologies myself so I am not sure if they are right for the company." Another owner-manager stated, "Most times I don't use these technologies because I am not good at using them. I have survived all of my life without using some of these things. The company has also been around for more than 20 years and it has lasted this long without these new gadgets."

Aside from the factor describes above, other variables came to the fore such as education, family status, and intellectual stimulation which influence leadership typologies as discussed in Chapter 7. At this juncture, establishing that leadership was the foremost element and that strategy was not the fundamental driver in the technological decisions of small owner-managed companies is imperative.

Strategic choices occurred as an effect rather than a cause for technology adoption. Owner-managers who made strategic decisions regarding the adoption of new innovations were typically leaders who introduce agile strategies. Contrastingly, inflexible leaders were naturally reluctant to strategy formulation—whether substantive or structural changes. Generally, owner-managers were found to be strategically weak due to no long-term planning for the companies. This may explain the disinterest in transformational technological tools like the internet which would demand a vast deviation from the norm.

Along with strategy, resources have been deemed as an internal influence on decision-making processes in companies.[12] Owner-managers identified the most essential resources to be financial and human. All the managers expressed resource constraints as such the challenge became one of budget allocation, human resource deployment, and priority spending. The small owner-managed companies in the sample had declining revenues due to the drastic reduction in commission from suppliers and therefore had to make critical deliberations regarding the allocation of limited resources. In light of limited resources, this study found that given the simple decision process and the impact of owner-managers personal interests, priority areas mirrored owner-managers preferences and interests. To substantiate this claim respondents stated the following, "The priority is paying bills, paying staff and keeping the business running from day to day. I have seen nothing to suggest that my limited finances must be spent on something just because it is new. I don't think I need it."

> I am not one of those who think that the use of the internet to reach clients is a waste of time but my profit margin is getting smaller and I have to prioritize how I use resources. My family depends on the survival of this business and I have to make sure that what I spend on will bring quick returns.

Companies may retain a relative position of competitiveness if they wield control of their resources in comparison to other firms without resources, if they act rationally.[13] Among the firms examined there was no significant disparity in their resource position given their declining

[12] Kraajenbrink et al. (2010).

[13] Penrose (1959), Colli (2011), Teece (1982), Wernerfelt (1984), Peteraf (1993), Mowery et al. (1998).

revenues because of the reduction in commission. The difference which arose was therefore not one of resource position instead it was of perception, among those firms that are technologically inclined, as distinguished in the earlier responses of low and high adopters. While there seemed to be no resource advantage for the firms in the present, in the future however, those companies that were investing in a greater online usage may acquire a resource position.

The resource position stalemate presents the basis upon which competing firms may outperform each other. Evident from these discussions is the significance of resources in a firm's decision-making process, it dictates what can or cannot be accomplished. While this may be true in absolute terms, resources are never infinite even for larger firms with a strong asset base. The decisions about how to use limited resources, is the defining measure. With the exception of two firms in the sample resource allocation to technology was not a priority area. The other firms who primarily used GDS system and were low adopters of technology had no desire to advance higher on the technological ladder. Notably, even those managers who recognized the importance of technological development insisted that there were other areas that demanded greater resources, namely staff renumeration and back-office functions. The prioritization of these areas was as a result of three key issues.

Firstly, the myopic vision of these leaders informed their thinking that on anything regarding the daily, immediate needs out weight strategic expenditure that would be realized in the future. This outlook may be a product of the leadership provided currently but more so in the last 5–10 years. One may infer that the shortsightedness exhibited by these firms have resigned them to their current survival mode, where markets and global industry have outpaced them. Thus, they are inept to see beyond the immediate crisis and are often left to allocate resources to play "catch up."

Secondly, leaders were unconvinced that there is a significant return on investments on technological investments. This may have shaped their mindset regarding technological investments as high risk, which also corroborates a finding which was presented previously in this chapter. Thirdly, which relates to competitiveness owner-managers were content with not investing in technological tools because their competitors locally were not investing either. For the managers in the sample this meant no major investment had to be made. Theorists posit that the available resources are closely linked to performance.[14] This is based on the notion that firms variety of internal capabilities and resources, directly informs its strategies through a meticulously organizational assessment.[15] Therefore, on the basis of heterogeneity its evident that strategic planning and decisions should be unique among firms and be dependent on the firm's asset base.

Although, the resource base of the firms examined was not identical, there was not sufficient heterogeneity to determine a relationship to differences in firm performance. The impact of the resource base of the firms was only distinguishable in terms of their decisions regarding areas to prioritize in terms of budget. The relationship between the company's leadership and the performance of these small, owner-managed firms was more pronounced. For instance, in terms of distribution transformation platforms local industry leaders were defined by specific characteristics.

Diversification whether of products, markets, or processes by companies is often pursued by those with broad resource bases.[16] Market and product diversification would stand to benefit greatly from increased use of the internet for sales and marketing in these companies. While increased resources would afford companies the luxury to decide on which platforms they may or may not capitalize on, the allocation of resources toward introducing advanced technology may serve as a catalyst in the diversification of other areas without further substantial investment. Given that a firm's resources is not the singular determining factor, they are conduits to many choices regarding strategy options as their internal capabilities allows for multiple avenues/opportunities for conducting business. Of the firms examined only one firm which is a market leader both technologically and financially demonstrated this concept. Their leadership position within the local industry was found to be attributed to diverging perceptions about the role of technology in the daily operations of the business. This business operational behavior embodies a broadening of their current understanding of resources, not limited to human and financial resources but also includes technological resources.

[14] Peteraf (1993).

[15] Freeman, R.E. 2010. *Strategic Management: A Stakeholder Approach*. Cambridge University Press.

[16] Montgomery and Hariharan (1991).

The inclusion of technological resources as a component of the asset base gives rise to an extended debate to include issues external to the firm. From the resource-based lenses, theorists[17] posit the utility of inter-firm collaboration. On the other hand Mowery et al. (1998) noted that technological overlap influence partner selection. The general applicability however infers that companies with significant resource base are more attractive for collaboration. A gap in the literature is the absence of which category of resources are most attractive and not much attention is paid to the differences among industries.

Within the travel industry small, owner-managed firms do not view collaboration with local competitors as a strategic tool. They cite the sharing of limited resources for this stance. There was also a protectionist belief among managers, in that believed they were slightly financially well of than their competitors and therefore safeguard this standing, from a strategic perspective. Respondents expressed that "any collaboration with other travel agencies would hurt my business because we are all struggling and some of these firms are even worse off than mine which would mean that they would suck the remaining life out of my business." "Experiences in the past have proven that many of the other local travel agencies cannot be trusted. Any alliance would mean the sharing of certain kinds of information and from a strategy point of view I don't like it. Since we all have challenges with money then we will each have to figure out how to survive."

At its core resource base discussions highlights a firm's heterogeneous capabilities, diversification strategy, and distinctive competencies.[18] Furthermore, distinctive competence is a function of the firm's resources at any given point. The longevity of this distinctiveness is dependent on the type of industry and resources. While the RBV intention is not to provide managerial prescriptions, it clearly highlights why some firms have a strategic advantage over others and its implications; therefore, it can be of significant value to managers.[19] A point critical to note however, is that most of the firms in the sample, did not have a distinct asset advantage. It was observed that strategies varied to some degree with high adopters and low adopters, and the consequent relationship between strategy and

[17] Teece (1982), Mowery et al. (1998).

[18] Mahoney and Pandian (1992).

[19] Kraajenbrink et al. (2010).

resources was identified as a crucial factor influencing strategy. High adopters included two firms exploring online selling options, one with a limited asset base and the other with a significant asset base which have proactive strategies that identified the need for them to create and invest in platforms that foreshadow changes in the market. Leaders of these firms were willing to allot resources to support these strategies. Typically, leaders of moderate and low adopter firms felt "local markets were not yet ready for this type of online activity" and were reactive to the changes in the market.

Based on the interviews conducted, it was gathered that lack of resources was the fundamental challenge facing most firms. One respondent opined: "You have to understand that we are a small company and we don't have a lot of money to go around." This proved to be a recurring theme from the sample, and the data gathered revealed that there was not a significant difference in the resource base of these firms. The firms were described as being reliant on external resources as opposed to internal capabilities for their functioning, which based on the data, revealed that most were operating at margins below variable costs. There was a difficulty in attaining specific data about financial resources of these firms to corroborate their pronouncements.

The study brought to the fore the diversity in how businesses chose to maximize the use of their limited resources. This was due in large to leadership decisions which are paramount given the small size and simple organizational structure, as evidence previously. A leader's actions are essential in resource allocation decision-making.[20] This is amplified in situations were firms have limited resources and must prioritize expenditure to ensure efficient use. Resource constraints primarily stated by owner-managers were human and financial resources. In many cases the managers bemoaned the lack of money to invest in high-tech strategy. Yet there were instances where managers felt they had no human resources capable of consistently maintaining technological innovations such as websites where money was available. This was bolstered by few companies which had websites but they were inactive. In light of resource limitations, technology adoption and training are not priority areas for investment based on leadership resistance. The owner-managers indicated that they were more inclined to use their limited resources in areas that would

[20] Kraajenbrink et al. (2010).

impact the daily operations than strategic objectives of which they were uncertain of the returns. Technological investments were therefore seen as risky investment and were not a priority area because it did not address the immediate needs of the market being served. On this basis leadership is a transactional process, that seeks to control daily activities and enforce procedures.

The scenario for the companies analyzed is quite unique in that participants are owner-managers who are invested in the immediate benefits of the business. This work purports that owner-managers are inherently transactional leaders, this is grounded on their desire to attain immediate returns on investments because it's the source of their family finances, despite the challenges of the industry. This postulation extends the arguments put forward in entrepreneurship literature, in that small business owners are deemed to be passive or reactive, risk averse, and non-inventive.[21] These traits are demonstrated by most of the owner-managers examined and is evident in their refusal to engaged in the inbound travel market due to an absence of commission even though it may yield long-term rewards. Carland et al. (2007) note that leaders of small businesses often operate the business as an extension of their individual personality and immediate needs.

To further explore these issues—short-term fatalistic goals, personal needs, and preferences owner-managers were asked to assess their technological needs and pair them with financial and human resources. Most placed emphasis on acquiring websites since they were low adopters of technology and have not reached this level on the hierarchical adoption ladder. Additionally, carefully crafted websites would facilitate online activity, which is ultimately being examined in this body of work. While this may be true the generally employees in these firms had low levels of formal education. Therefore, their experience with learning to use new technologies was described by managers as a tedious process, as many found it difficult. The typical employee only acquired a high school or 3 GCE O'level subjects or the Caribbean Secondary Examination certificate. In comparison to other front-line workers such as those in the banking industry which required the minimum of 5 subjects with preference given to persons with a Bachelor's degree these qualifications of the employees in these small owner-managed firms were low. It therefore

[21] Covin and Slevin (1988).

becomes problematic because these firms hire the cheapest labor and then fail to add value, given that they lack the resources to facilitate general staff training it is unsurprising that technology training was non-existent. Given the resistance of owner-managers to adopt some types of technology, they are even more discouraged by the possibility of providing training for employees who may be difficult to train. Added to this, the employees within these firms are less likely to be creative and critical thinkers and may enjoy the routine and monotony of work. The idea is that even if administrators were intellectually stimulating, workers who are more satisfied with pursuing orders would give very little reply. Directly linked to financial resources of small owner-managed firms are all aspects of human resources such as recruitment, training and selection, managers stated. As a result, all the respondents noted an increase financial resource base would see a greater allocation toward training of their human resources; however, a few noted that an increased financial resource base would not guarantee greater investment in technology adoption within the short to medium term. Analysis of the firm's financial asset will now be done.

The declaration by owner-managers that the lack of financial resources negatively impact technological adoption suggests that they have a desire to adopt but is hindered by the scarcity of resources. Yet, this is incongruent with the subsequent revelation where most owner-managers expressed there was no other technology which they wanted to invest. This general unwillingness to adopt is more evident than the inability to adopt. Having said this the financial challenges of the firms must not be overlooked even though it is not direct driver of technological adoption, its influence is indirect. The lack of adequate financial resource base is accompanied by companies being in crisis management mode which doesn't facilitate long-term strategic planning. This reality is emblematic of two firms whose livelihood is dependent on the business survival and profitability. This presupposes the over-arching catalyst as leadership and the factors that affect each person, rendering him/her unique in the local industry from their counterparts.

A firm's behavior is impacted by its internal factor—resources[22] and strategy.[23] The behavior under investigation in this study is technology

[22] Barney (2001), Kraajenbrink et al. (2010).

[23] David (2007), Stonehouse and Snowdon (2007).

adoption of small owner-managed firms specifically for online sales and marketing activities. Through primary and secondary data, multiple variables such as strategy, resources, digital divide, culture, and leadership which affect decision-making were examined. Distillation of the findings from this analysis indicated that strategy and resources were no the key catalyst of technology adoption. Yet, they should not be dismissed as they act as peripheral factors in that managers employ them to affect technology adoption process but is not the origin for these processes. These peripheral contributors arise a by-product of leadership style rather than drivers in their own right. These variables may have a profound impact in a different context but within an information intensive environment, specifically small owner-managed firms they are mere pawns in technology decisions.

Firm resources and strategy have been described in the literature as relevant to the debate on firm behavior, as shown in Fig. 2.1. These factors, together with leadership, culture, and the digital divide, were also part of the organizational decision-making theory in Fig. 2.2. This method was employed to identify the contributors of technology adoption in small owner-managed travel companies along a hierarchy. Exploring these internal factors (strategy and resources) further through primary data collection with several owner-managed travel companies in Jamaica showed that there was no clear strategy influencing the conduct of technology adoption. It could be concluded that many companies were ultimately engaging in a hi-touch approach (intense personal interaction) and worried that they might experience the potential psychological threat of reliance on technology as presented by Eastlick and Lotz (1999). In addition, the owner-managers perceived a hi-tech and hi-touch strategy was incompatible.

Owner-managers have demonstrated from a management standpoint that they face limitations on human and financial capital. They argued that this hindered their ability to adopt higher levels of innovation was archaic, as they expressed no genuine desire to adopt new innovations, resource constraints, or not. Consequently, decisions about how to distribute even limited resources are now becoming owner-managers sole remit. The tendency to prioritize spending in areas other than adoption of technology was a reflection of the preferences and comfort zone of the individual leader.

Ultimately, firm strategy and resources have not adequately explained differences in company adoption behavior. However, such considerations are helpful in recognizing how resources were used to make innovation choices in the overall sense, but they were less relevant in describing why some companies were at different levels of the technology adoption hierarchy, provided that policies and resource constraints were not heterogeneous enough to answer the question of why some firms were more technologically advanced than others. These internal factors arose as a consequence of the disparities in management of these independent, owner-managed travel companies rather than as major contributors to the decision to adopt cutting-edge technology. In view of the inability of the internal micro factors to clarify the adoption activity of these companies, more analysis and examination of whether or not macro factors, such as society and the virtual divide, are best able to explain business adoption behavior, is necessary.

References

Barney, J.B. 2001. "Resource-Based Theories of Competitive Advantage: A Ten Year Retrospective on the Resource-Based View." *Journal of Management* 27: 643.

Bethapudi, A. 2013. "The Role of ICT in Tourism Industry." *Journal of Applied Economics and Business* 1(4): 67–79.

Buhalis, Dimitrios, and Maria Cristina Licata. 2002. "The Future eTourism Intermediaries." *Journal of Tourism Management* 23(3): 207–220.

Carland, J.W., F. Hoy, W.R. Boulton, and J.A.C. Carland. 2007. "Differentiating Entrepreneurs from Small Business Owners: A Conceptualization." In *Entrepreneurship*, 73–81. Berlin and Heidelberg: Springer.

Colli, A. 2011. "Business History in Family Business Studies: From Neglect to Cooperation? *Journal of Family Business Management* 1: 12.

Covin, Jeffery G., and Dennis Slevin. 1988. "The Influence of Organization Structure on the Utility of an Entrepreneurial Top Management Style." *Journal of Management Studies* 25(3): 217–234.

David, Fred. 2007. *Strategic Management: Concepts and Cases*. Upper Saddle River, NJ: Prentice Hall.

Davis, Fred D. 1989. "Perceived Ease of Use, and User Acceptance of Information Technology." *MIS Quarterly* 13(3): 319–339.

Eames, Donald, and Gregory Norkus. 1988. "Developing Your Procurement Strategy." *Cornell Hotel and Restaurant Administration Quarterly* 29: 30–33.

Eastlick, Mary Ann, and Sherry Lotz. 1999. "Profiling Potential Adopters and Non-Adopters of an Interactive Electronic Shopping Medium." *International Journal of Retail and Distribution Management* 27(6): 209–223.

Freeman, R.E. 2010. *Strategic Management: A Stakeholder Approach*. Cambridge: Cambridge University Press.

Hatton, Mike. 2004. "Redefining the Relationships: The Future of Travel Agencies and the Global Agency Contract in a Changing Distribution System." *Journal of Vacation Marketing* 10: 101.

James, Jeffrey. 2004. "The Global Digital Divide in the Internet: Developed Countries Constructs and Third World Realities." *Journal of Information Science* 31(2): 114–123.

Kim, Eonsoo, Nam Dae-Il, and J.L. Stimpert. 2004. "The Applicability of Porter's Generic Strategies in the Digital Age: Assumptions, Conjectures, and Suggestions." *Journal of Management* 30(5): 569–589.

Kraajenbrink, Jeroen, J.C. Spender, and Aard Groen. 2010. "The Resource-Based View: A Review and Assessment of Its Critiques." *Journal of Management* 36: 349.

Lynskey, Michael. 2004. "Determinants of Innovative Activity in Japanese Technology-Based Start-up Firms." *International Small Business Journal* 22(2): 159–196.

Mahoney, Joseph T., and J. Rajendran Pandian. 1992. "The Resource-Based View Within the Conversation of Strategic Management." *Strategic Management Journal* 13: 363–380.

Montgomery, Cynthia A., and Sam Harihan. 1991. "Diversified Expansion by Large Established Firms." *Journal of Economic Behaviour* 52(1): 71–89.

Mowery, David C., Joanne E. Oxley, and Brian Silverman. 1998. "Technological Overlap and Interfirm Cooperation: Implications for the Resource-Based View of the Firm." *Research Policy* 27: 507–523.

O'Connor, Peter. 2008, January. "User-Generated Content and Travel: A Case Study on Tripadvisor.com." In *ENTER*, vol. 2008, 47–58.

Penrose, Edith T. 1959. *The Theory of the Growth of the Firm*. New York: Wiley.

Peteraf, Margaret A. 1993. "The Cornerstones of Competitive Advantage: A Resource-Based View." *Strategic Management Journal* 14: 179–191.

Poon, Auliana. 1993. *Tourism, Technology and Competitive Strategies*. Wallingford: CAB International.

Porter, Michael. 1985. *Competitive Advantage*. New York: The Free Press.

———. 2001. "Bewährte Strategien Werden mit dem Internet Noch Wirksamer." *Harvard Business Manager* 5: 64–81.

Reve, Torger. 1990. "The Firm as a Nexus of Internal and External Contracts." In *The Firm as a Nexus of Treaties*, edited by Masahiko Aioki, 136–188. London: Sage.

Ricketts, Martin. 1994. *The Economies of Business Enterprise*. Hempstead: Harvester Wheatsheaf.

Rogers, Everett. 2003. *Diffusion of Innovations*. 5th ed. New York: The Free Press.

Ryals, Lynette, and Andre Humphries. 2007. "Managing Key Business-to-Business Relationships: What Marketing Can Learn from Supply Chain Management." *Journal of Service Research* 9: 312.

Stonehouse, George, and Brian Snowdon. 2007. "Competitive Advantage Revisited: Michael Porter on Strategy and Competitiveness." *Journal of Management Inquiry* 16(3): 256–273.

Stump, Rodney L., Wen Gong, and Zhan Li. 2008. "Exploring the Digital Divide in Mobile-phone Adoption Levels across Countries: Do Population Socioeconomic Traits Operate in the Same Manner as Their Individual-Level Demographic Counterparts?" *Journal of Macromarketing* 28: 397–412.

Tan, S.J. 1999. "Strategies for Reducing Consumers' Risk Aversion in Internet Shopping." *Journal of Consumer Marketing*.

Teece, David J. 1982. "Toward an Economic Theory of the Multiproduct Firm." *Journal of Economic Behaviour and Organization* 3: 39–63.

Wernerfelt, Birger. 1984. "A Resource-Based View of the Firm." *Strategic Management Journal* 5: 171–180.

Williamson, Oliver E. 1979. "Transaction Cost-Economies: The Governance of Contractual Relations." *Journal of Law and Economics* 22: 232–261.

———. 1990. "The Firm as a Nexus of Treaties: An Introduction." In *The Firm as a Nexus of Treaties*, edited by Masahiko Aioki, 136–188. London: Sage.

Wöber, Karl, and Ulrike Gretzel. 2000. "Tourism Managers' Adoption of Marketing Decision Support Systems." *Journal of Travel Research* 39: 172–181.

World Travel and Tourism Council. 2019. *Travel & Tourism-Economic Impact 2019-World*.

Wynne, Clive, Pierre Berthon, Leyland Pitt, Michael Ewing, and Julie Napoli. 2001. "The Impact of the Internet on the Distribution Value Chain: The Case of the South African Tourism Industry." *International Marketing Review* 18(4): 420–431.

CHAPTER 4

External Factors: The Digital Divide, Closing the Gap

An important area of focus in the literature on Organizational Theory is culture, which embraces both national and organizational culture. In this study, emphasis will be placed on national culture. This is as a result of the analysis of small, owner-managed firms, in the preceding findings chapter which established that there is a link between internal factors of the organization and their malleability by the leaders of these firms whose business decisions and practices often reflect their preferences and experience. Considering this, internal factors such as structure, organizational culture, resources, and strategy are recognized as secondary factors in influencing firm behavior in small owner-managed firms. This research conceptualizes national culture in terms of values, norms, traditions, and social interactions within the society. National culture therefore comprised of external factors that may cause a particular effect.

Hofstede (1980) is widely consulted as an authority on international cultural differences in the workplace. Nonetheless, McSweeney (2002) critiqued Hofstede's position by arguing that national culture rests on plurality and therefore, any work that assumes uniformity within national cultures is flawed. This plurality he believed, constituted the observable differences in culture. McSweeney questioned whether culture can systematically cause differences in behavior among people from various countries. The findings of this research indicate that within the same national culture there were differences in the technological adoption behavior among firms. This underpinned the exploration of specific

A. Spencer, *Technology Adoption in the Caribbean Tourism Industry*,
https://doi.org/10.1007/978-3-030-61584-0_4

cultural paradigms that influenced behavior especially adoption behavior in firms.

From the customer's perspective, societal values are influential. Owner-managers often expressed that "the most important and valuable thing to customers is the personal touch." This meant that for most respondents' owner-managers should align their efforts by providing incomparable one-on-one experiences rather than searching for the most efficient distribution method. Furthermore, given their perception about online sale providing minimal value to the customers this has influenced their belief that the introduction of some innovations was useless. One respondent noted, "we could use all of these fancy technologies but if our customers don't see it as something that is necessary then we would all be wasting our time trying to do something just because it's fashionable."

It is clear that customer's values influenced the firms values considerably, the aforementioned statement is a reflection of the majority of the firm's beliefs. Yet there was difference in views between high and low adopters. The two firms at the top of the hierarchical ladder stated:

> While a large percentage of the Jamaican population does not place a high value on e-commerce activities today we recognize that those values are changing fast and we can even see this in the wide acceptance and use of smart phones such as blackberry.
>
> We have to be prepared for market changes which are on the horizon. Even if we only continue to serve this local market, there is a revolution taking place with our young people which we must anticipate.

Highlighting the differences is useful considering that owner-managers have been exposed to a similar national culture. The differences in perception indicate that there are other key factors external to the value construct which influence the different adoption behavior among the firms. These differences are useful for academic consideration in so far as uniformed differences exist in national cultures. Exploration the impact of national culture on the organization revealed a significant correlation between organizational behavior and elements of national culture like norms and traditions.[1]

Societal norms and traditions represent the way in which people do and perceive things and are critical in overall cultural influence.

[1] Gelade et al. (2006).

While this was evident in the local clientele, it was more pronounced among the firms and is conveyed by sentiments such as, “we have been doing it this way for over 20 years and it has worked for us. If we change our approach we do not know if it will work just as well. Also most of us in this company are not comfortable with changing how we make bookings.” As the respondents were noticeably traditionalist in their approach to business management, perhaps they were suffering from what Lang (2000) considered as “techno fear.” The two are not discrete in the sense that it is a subscription to traditions, customs, and norms that require a rejection of anything that contravenes the established “normal” way of doing things. Critically, it was found that high adopters held different perceptions regarding traditions. The most technological savvy firm leader articulated:

> “Nothing remains the same forever. There was a time when we did manual bookings and we did not have any GDS. We learned to use the GDS and today many of us cannot do without it. In the same way we always have to be ready to adjust based on new developments.” The divergence in views emphasizes that while national traditions and norms impact the collective views of the people, there are other elements that must provide an explanation as to the varying views and adoption behavior among firms. Significantly, it is important to point out that all the owner-managers have resided in Jamaica all their lives, external cultural influences can be considered limited except for short stay travel and media exposure.

The emergence of an indirect cultural influence was brought to the fore as owner-managers expressed that a type of relationship-oriented society existed, whereby customers were shaped by culture thus making it virtually impossible for them to ignore customers’ preferences in their business practices. This relationship orientation is indicative of the perception that high regard is placed on personal interaction by customers in the local outbound market which could not be replicated online. Thus, issues of marketing and distribution strategy are brought to the fore; but at the same time, leadership constraints are revealed, primarily those that impact decision-making capacities. Respondents stated that:

> I would love to see the internet listen to their complaints and empathize and ask how their family is doing. These are the things which my customers love because Jamaicans are very friendly people and I have a good relationship will all of my customers.

> A lot of my business comes through a major company that I used to work for because I still have a good relationship with them and their families. They always book with me no matter what and we are good friends even before we are service provider and customer.

> Someone in Jamaica always knows someone else. Once you form a relationship with a client they always tell someone else how warm and friendly you are. Before you know it you have a new client because they recommended you and that person wants the same personalized treatment.

The relationship-oriented society negatively impacts technological adoption in firms, most respondents were assured that providing service in a relationship-oriented society characterized by personal interaction which is valued above efficiency and speed, would safeguard their future business. Contrastingly, high adopters were less inclined to take this stance. It was highlighted by the market leader that:

> A number of things are changing. First of all our loyal customers are not as loyal anymore, so the relationships seem to mean less to them now. Secondly these loyal clients are getting old and are traveling less so we have to focus on the new generation and they are all about the best price and convenience. And don't forget that they are internet savvy. As I mentioned to you before they do everything on their mobile smart phones.

Comparable to the previous constructs of traditions, values, and norms, the relationship-oriented culture facilitates the low technology adoption among firms however, it provides no explanatory power as to why minority of firms diverge from the norms and endeavor to be more involved in innovative behavior and more proactive. To understand the differences in innovative behavior, a closer look is required, particularly at the association between culture and innovation. From the primary data collection and analysis, it was revealed that national cultural circumstances usually shaped individuals' views about technological innovations in response to what is observed in the society as leaders admitted. Therefore, a relatively low level of demonstrated technological activity in the country would result in the respondents' inability to readily identify the benefits and uses of the technology. In this vein, one respondent stated that:

> All of these things that you mention are things that I hardly see anyone use. I don't think that Jamaica is a country that uses a lot of technology. I know that the latest craze is Facebook but young people only use it to talk to friends. I can't say that I have seen any evidence to suggest that we are high technology users.

The challenge is that this response does not offer any explanation for the differences in adoption behavior, although the national culture was common in the firms that were sampled. A deeper examination and analysis indicated that the dominant subcultures to which these individuals belonged was dominant in molding their views about technology use. Emerging as the most dominant subculture influencing perceptions was the family.

Subcultures possess their own traditions, values, and norms. The family is one such subculture and is a primary source of socialization.[2] This suggests that some individuals might be more inclined to engage in some behaviors that are of their subculture rather than of the broader national culture. Since leadership has been established in the previous chapters of this research as a critical determinant of what new technologies were to be adopted in small, owner-managed firms, it is imperative to identify the factors that influence the growth of these leaders. From a cultural perspective, the respondents' family backgrounds played a role in so far as there was innovative behavior occurring in the families. Leaders from families with children who were technologically savvy, were seemingly more inclined to engage in those behaviors personally, this translated into their business activities. One owner-manager expressed that "I am very open to using the technology personally because I have seen my son buy all these things online using my credit card. As a teenager he has taught me how to buy things online too. I realize that if I can buy online then why not sell online. This is why I am now exploring the option." This emanated from having a great level of comfort utilizing tools which they observed a trusted family member using.

[2] Haralambos and Holborn (2008).

Culture and Leadership

This research is a complement to the work done by Elenkov and Manev (2005) as it has found that cultural context shapes leadership and regulates its relationship with organizational innovation. However, there are some differences in terms of the importance given to cultural influence and the point of emphasis. This research has found that cultural influence, in contrast to the postulations of the above authors is more indirect. While a cultural influence was detected in the sample, there were some significant dissimilarities in behavior that national culture could not explain since all owner-managers had this commonality. Perhaps, a comparative study with Asian (Jung et al. 2002) or European (Elenkov and Manev 2005) nations may expose several differences due to national culture however within the same culture there is less resonance.

Additionally, while the family was discovered as the most influential subculture to impact innovative behavior, it was still insufficient to totally explain adoption behavior in the firms. Similarities in family composition were shared across some firms at higher levels of adoption and some firms at lower levels of adoption. While the evidence indicated that there was consensus among leaders about their families being influential in their degree of comfort introducing technologies in their firms, still there were variances in the levels of adoption. An overall influence was evident in the differences between low adopters and high adopters but this was less distinct between moderate adopters and high adopters who shared similarities in how their families were composed. Thus, it was shown that alternative explanations were required to clarify the differences in firm behavior. This could be done through an exploration of other factors in combination with family. Possibly, these factors could provide an explanation of leadership differences and ultimately the differences surrounding technology adoption.

Singh and Krishnan (2007) contemporary work outlined that cultural sensitivity is required when addressing transactional and transformational leadership in the leadership literature. Considering this it is vital to validate a relationship between leadership style and culture and to also determine while leadership is perceived as a critical driver in technology adoption in this research, broader considerations from a *Critical Social Science* perspective and its role in a subculture such as family must be addressed.

Interestingly, respondents highlighted that their competitors were molded by the national culture, this reduced the necessity for adopting new technologies since the competition was not doing so. Each respondent echoed this sentiment about the other respondents. While it appears that there was no perceived cultural influence on their own leadership style or the inclination to adopt technology innovations, the converse was stated to be an overwhelming consensus. However, this gave rise to a challenge, in that it provided no explanatory power for two things.

Firstly, it does not explain whether culture has an effect on leadership characteristics as conceptualized in the literature on transformational leadership or whether it simply influences the perceptions of technology. Understanding this distinction is necessary since the effect of strong transformational leadership characteristics should enable a leader to surrender his biases where there is adequate evidence of the need for transformation. Secondly, it provides no explanation as to why there are divergent views within the same culture about technology emerged. The cultural influence is therefore treated as an indirect one, as stated earlier, which affects leadership peripherally and does so in combination with other factors.

The comprehension of culture is interrelated to the wider sociological discourse which outlines customs, norms, and values as essential constructs in the discussion of a social system. Notably, classical theorists of technological adoption like Davis (1989) and Rogers (1962) acknowledged and outlined the significant influence of the social system on adoption. The uniqueness of each social system based on their characteristics differentiate it from other systems.

These cultural variances affect work-related innovation, behavior, and values. Ultimately, the leader was perceived as a product of cultural circumstance. These observations warranted a deeper investigation of culture as a determining factor; however, the primary study found that the influence was indirect. Within the sample cultural influence was recognized but some key differences in behavior could not be explained by national culture since it was shared among owner-managers. The influence of the family was perhaps more insightful since, as a subculture, it could explain large disparities among the leaders yet it still did not offer an explanation for subtle differences between leadership categories. Owing to this, the researcher was compelled to examine other factors that could offer an explanation. The digital divide may facilitate a much deeper discussion when framed in a global context and within societies.

Any apparent dissimilarities within the society may facilitate the explanation of divergent technology adoption of firms.

The Digital Divide

Upon the introduction of local telecommunications market, new submarine cables were laid that connected Jamaica, The Bahamas, and Trinidad and Tobabgo to the United States. The implementation of these new cables increased the collective number of submarine cable connecting these countries to the world to four. While broadband digital penetration was much slower than in developed countries such as the United States and the United Kingdom, connectivity levels were still high as a result of smartphones use. In support of Stump et al. (2008) work on mobile technologies which outlines there has been rapid diffusion throughout developing countries which has reduced the disparity between developed and developing nations. Nonetheless, a global digital divide is evident as increased technology infrastructure has not translated into the increased access within the average home. One of the respondents highlighted this by stating that:

> It would be a waste of time to try to reach our customers online because many of them do not even have the internet at home. Even though the younger ones are using smart phones most of them are not buying anything online with their phones at all.

The respondents were somewhat satisfied with the technology infrastructure on the island and mentioned that it was provided at a reasonable cost. However, in their view, many Jamaicans were simply not interested. "The technology is available but I just don't think we are that kind of society. We thrive on personal contact." This evidence corroborates with the annual e-Readiness survey that is conducted routinely, and compiled and published by the Economist Intelligence Unit (EIU) which is the business information arm of the Economist Group; publisher of The Economist magazine. In Jamaica's third appearance in the 2007 EIU report Jamaica's overall score improved marginally from 5.03 to 5.05 but Jamaica's ranking fell three places from 43 to 46. Trinidad appeared in the report several years later and ranked below Jamaica while the Bahamas was not a participant.

Broadband affordability and penetration have now garnered much attention in the survey. According to past data, Jamaica ranking in the area of telephone and narrowband penetration was exemplary. However, their performance in the area of Connectivity and Technology Infrastructure declined with the removal of the above category. Additionally, Jamaica's score decreased in the Consumer and Business Adoption category due to the revamping of the criteria in this group which would assess the number and level of online commerce and government services rather than the availability of finance and IT personnel. Of particular interest online commerce for travel services is virtually non-existent, this therefore needs to be addressed. This however is not an issue of access and as such the access divide purported by the theorists above fails to explain the adoption levels in firms. The digital divide which was conceptualized in Chapter 1 outlines there are also instances of unequal access to technology within societies. The evolution theory on the digital divide highlights issues of access stemming from inequalities that are economically driven,[3] politically driven,[4] and socially driven.[5] However, while the perception of economically driven inequality was revealed by the data, this did not differentiate high technology adopters from low technology adopters. Nonetheless, access to technological resources, while less when compared to developed countries, was equal within the business landscape for Jamaican companies. James (2004) postulations indicate that increasing access reconciles the gap between users. He argues that there exists innovative utilization of technologies in some developing countries despite individual access being limited in some countries. This leaves Rogers (2003) position unchallenged which posits that significant focus has been given to the access divide, marginalizing the content divide, and learning divide which will present a disadvantage for some. The key insight provided by this argument is the issue of aligning content to the audience needs.

A rational as to the differences in firm behavior is provided by the content and learning divides but there were other contributory factors. For example, slight differences were found in terms of the learning divide, which reflected the skills and competencies of the employees. It was

[3] Yu (2006).

[4] Feather (1998).

[5] Lash (1994).

concluded that this is a human resource issue and thus, would be more suited for resource-based discussions. The content divide appeared to be a secondary factor as western culture preferences shaped the majority of content meanwhile due to its close proximity to the United Sated the Caribbean benefitted. It is most important that ideas of the digital divide be relevant to the industries. Proponents of the global digital divide[6] generally posit that economically developed nations control the development of ICT and access which automatically gives them a distinct advantage. However, this body of work provides minimal specifications as to the solutions[7] to circumvent the automatic advantage created by economically developed countries, while majority of the earlier research paid a lot of attention to the information inequality within societies, recently there have been more focus on the global digital divide which addresses the dissimilarities across nations. This shift has accommodated a more rigorous debate within the travel and tourism discourse where supply and demand issues are explored.

Maurer and Lutz (2011) support the postulations of Minghetti and Buhalis (2010) by theorizing that there is a substantial communication gap within travel and tourism industry between the supply and demand sides. They argue that there are four categories of communication gaps: the access gap, usage gap, skills gap, and attitude gap. They convincingly argue that these gaps occur on a hierarchy which means that one level has to be completed before the next.

While this argument provides no explanation of the differences in firm behavior in this study it was necessary to examine these issues on the basis that there are industry implications. The researcher contends that irrespective of the dramatic decrease in retail travel firms in the last two years, they still have the potential to fill a very big gap in the industry. The Jamaica Tourism Board, The National Tourism Office has been unsuccessful in closing the digital communication gap with developed countries, therefore the opportunity exists for outbound agents to increase their engagement in the inbound market by increasing the online communication between countries. An understanding of the current gaps is necessary if the re-engineering of their functions is to occur. The gaps between the

[6] Mosaic Group (1998), Castells (2000), Norris (2000), Rogers (2003), Kirkman et al. (2002), Drori and Jang (2003).

[7] James (2004).

domestic firms and the developed markets which demand tourism services were revealed by the interviews to be similar to the four areas highlighted earlier: access gap, usage gap, skills gap, and attitude gap. However, while a uniformed hierarchical sequence was earlier purported by Maurer and Lutz (2011) this work did not find this uniformity of sequence.

The data exposed that access was a prerequisite for usage however for the upper tiers of the hierarchy this did not hold true. In many cases the existence of an attitude gap and skills gap occurred simultaneously while in other instances the skills gaps seemed to be higher on the hierarchy when compared to the attitude gap. This research findings debased the notion that attitude always sits at the apex of the multiple digital divide pyramid. Within the context of this research, digital divide was conceptualized in terms of political, social, or economic factors and whether they were determinants of information inequality. Comparatively, economic factors were the most dominant in terms of access to technological resources globally. Since Jamaica experienced phenomenal levels of mobile smartphone penetration, this gap has been reduced substantially. Nonetheless, a communication gap persists between the destination and developed high-access nations, from which the country obtains international tourists.

More importantly, small owner-managed firms had equal access to technological resources within the society. From the assessment it was clear that digital divide offered no explanation as to the differences in firms behavior in the society. Its only utility to this study is its ability to inform overall industry challenges for the small owner-managed firms. In the face of these common challenges, all the firms sampled still behave differently in terms of technology adoption and as such these differences necessitate alternative explanations. Context issues like culture and digital divide provides an explanation of the industry's slow IT uptake. Yet there was no evidence of access inequality or significantly different cultural practices among firms thus these variables could not explain adoption differences between firms.

Contextual issues such as culture and the digital divide facilitate an elucidation of the industry's behavior overall, but they are insufficient when considered to explain the difference among firms. Therefore, it is still not evident why these firms implemented different operational and strategic choices given that they exist to operate in analogous environment. The assessment of culture as factor exposed the relationship-oriented society in which firm operated where familiarity and friendships influenced business relationships. In many instances, oral advertisement from existing clientele who became “friends of the business” was one method by which firms attracted clients. This impacted perceptions of whether increased technology adoption was necessary in two ways. Firstly, some respondents were of the view that their client base was stable and secure and it required little effort to encourage their loyalty. Secondly, they stated were very familiar with their clients and reasoned that their relationship orientation conveyed that they were culturally inclined to requiring intense personal interaction, which could not be provided by the internet.

In addition to a relationship orientation, the impact of the national culture was the promotion of traditionalism with strict obedience to norms within the society. This impacted both firm and customer behavior and influenced strategic decisions which required a deviation from established norms. At the national level all participants of the study experienced these cultural factors which indicate that they are not suitable to explain the individual differences regarding firm adoption behavior. A more influential element of culture arose from the subcultural influence of the family.

Through its influence on leaders in small owner-managed firms family composition played an integral role in the firms’ technology adoption. While employee’s family background may have shaped their perceptions, it was deemed more pertinent to examine its influence on leadership within small owner-managed firms given the leaders unilateral authority to make decision about technology adoption and innovation. Family composition was found to be significant especially in distinguishing high adopters (those in the initiation phase of online selling and the use of social media as a marketing tool) from low adopters.

While the family background of employees may also have shaped their views it was deemed to be more important to assess its influence on leadership in the context of small owner-managed firms where leaders make unilateral decisions about innovation. When it comes to separating high adopters (those in the initiation phase of online selling and the use of social media as a marketing tool) from low adopters, family composition was found to be significant.

The evidence revealed that those owner-managers who shared the same household with their families were more open to innovative ideas and were more inclined to introduce innovation in the firms. In particular, teenagers who engaged in online shopping made a big difference as respondents saw the benefits of operating in cyberspace. Even though the influence of the family as a subculture elucidated larger disparities, this subculture offered no explanation for subtle nuances between leadership categories. For instance, there were high adopters and moderate adopters who had the same family composition. This demonstrates that used in isolation this variable is not explanatory and requires the combination with other variables to elucidate variations in the firms' technology adoption behavior. The digital divide was also a less significant context factor used to explain firm adoption behavior. The firms sampled for the study had equal access to technological resources and encountered similar challenges regarding content and learning divides. The information inequality driven by social, political, and economic factors within the society resonated more with the wider population that with these middle-class business owners.

The usefulness of the digital divide debate becomes more apparent when it is applied to the travel and tourism industry as a destination imperative as opposed to a firm imperative. There is more of a global divide than one which is internal to the country in the business context. After bridging the technological divide[8] and communication gap[9] that exits small owner-managed travel firm may rebrand themselves as being relevant. The firm's reinvention may be achieved by using covertly induced and autonomous agents such as the internet.[10] The explanatory potential of the digital divide and the national culture in understanding technological adoption differences in the firms sampled are weakened due to the level playing which exists for both factors within the country. However, leadership emerged as the most significant factor in clarifying technological adoption behavior at different level for small owner-managed firms with equal access and opportunity to technological innovation.

It has been demonstrated that leadership is the essential determinant of technological adoption behavior in small owner-managed travel firms. Factors such as firm strategy and resource allocation were manipulated to meet the needs of the owner mangers and were more of an effect than a causal factor in the technology adoption discourse for these types of

[8] Minghetti and Buhalis (2010).

[9] Maurer and Lutz (2011).

[10] Govers et al. (2007).

firms. Macro factors like culture and the digital divide were more useful in explaining why the travel industry as a whole IT uptake was slower than industries in other countries. Given the even influence of both these factors on the firms within the society, the extent of their use is instructive in providing enlightened strategy regarding the industry restructuring. Primarily, this study aims to identify what factors influence small owner-managed travel firms adoption decisions and the key factor that influences is leadership.

These firms are seen as extensions of their individual leaders which are the owners of these firms and make all decisions whether minor or major. Typically, these decisions reflect their individual preferences and experiences.

The following chapter provides a comprehensive discussion of the leadership element and what variables influence greater technological adoption behavior in these firms. Ultimately, these leadership characteristics are combined to form leadership typologies that influence adoption along a hierarchy.

References

Castells, Manuel. 2000. *End of Millennium*. 2nd ed. Oxford: Blackwell.

Davis, Fred D. 1989. "Perceived Ease of Use, and User Acceptance of Information Technology." *MIS Quarterly* 13(3): 319–339.

Drori, Gili, and Yong Suk Jang. 2003. "The Global Digital Divide: A Sociological Assessment of Trends and Causes." *Social Science Computer Review* 21: 144–161.

Elenkov, Detelin, and Ivan Manev. 2005. "Top Management Leadership and Influence on Innovation: The Role of Socio-cultural Context." *Journal of Management* 31(3): 381–402.

Feather, John. 1998. *The Information Society: A Study of Continuity and Change*. 2nd ed. London: Library Association Publishing.

Gelade, Garry, Paul Dobson, and Patrick Gilbert. 2006. "National Differences in Organizational Commitment: Effect of Economy, Product of Personality, or Consequence of Culture?" *Journal of Cross-Cultural Psychology* 37(5): 542–556.

Govers, Robert F., Frank M. Go, and Kumar Kuldeep. 2007. "Promoting Tourism Destination Image." *Journal of Travel Research* 46(1): 15–23.

Haralambos, M., and M. Holborn. 2008. *Sociology: Themes and Perspectives*. UK: HarperCollins.

Hofstede, Geert. 1980. *Culture's Consequences: International Differences in Work-related Values*. Beverly Hills, CA: Sage.

James, Jeffrey. 2004. "The Global Digital Divide in the Internet: Developed Countries Constructs and Third World Realities." *Journal of Information Science* 31(2): 114–123.

Jung, Timothy H., and Richard Butler. 2000. "The Measurement of the Marketing Effectiveness of the Internet in the Tourism and Hospitality Industry." In *Information and Communication Technologies in Tourism 2000: Proceedings of the International Conference in Barcelona, Spain*, edited by Daniel Fesenmaier, Stefan Klein, and Dimitrios Buhalis, 460–472. Wien: Springer.

Jung, Timothy H., Panos Louviers, and Harmen Oppewal. 2002. "Channel Management Strategy in the E-commerce Environment: A Portfolio Management Approach." In *Information and Communications Technologies in Tourism*, edited by Karl Wober, Andrew Frew, and Martin Hitz, 17–26. New York: Springer Verlag.

Kirkman, Geoffrey, Peter Cornelius, Jeffrey Sachs, and Klaus Schwab. 2002. *Global Information Technology Report 2001–2002*. Oxford: Oxford University Press.

Lang, Tania. 2000. "The Effect of the Internet on Travel Consumer Purchasing Behaviour and Implications for Travel Agencies." *Journal of Vacation Marketing* 6: 368.

Lash, Scott. 1994. "Reflexivity and its Doubles: Structure, Aesthetics, Community." In *Reflexive Modernization*, edited by Ulrich Beck, Anthony Giddens, and Scott Lash, 110–173. Cambridge: Polity Press.

Maurer, Christian, and Veronika Lutz. 2011. "The Impact of Digital Divide on Global Tourism: Strategic Implications of Overcoming Communication Gaps caused by Digital Inequalities." In *Information and Communication Technologies in Tourism 2011 Enter Proceedings*, edited by Rob Law, Matthias Fuchs, and Francesco Ricci, 265–277. New York: Springer Wein.

McSweeney, Brendan. 2002. "Hofstede's Model of National Cultural Differences and the Consequences: A Triumph of Faith—A Failure of Analysis." *Human Relations* 55: 89–118.

Minghetti, Valeria, and Dimitrios Buhalis. 2010. "Digital Divide in Tourism." *Journal of Travel Research* 49(3): 267–281.

Mosaic Group. 1998. "The Global Diffusion of the Internet Project: An Initial Inductive Study." Accessed January 16, 2010. www.agsd.com/gdi97/gdi97.html.

Norris, Pipps. 2000. "Digital Divide? Civic Engagement, Information Poverty and the Internet in Democratic Societies." Accessed November 9, 2010.

Rogers, Everett. 1962. *Diffusion of Innovations*. New York: The Free Press.

Rogers, Everett. 2003. *Diffusion of Innovations*. 5th ed. New York: The Free Press.

Singh, Niti, and R. Venkat Krishnan. 2007. "Transformational Leadership in India: Developing and Validating a New Scale." *International Journal of Cross Cultural Management* 7: 219.

Stump, Rodney L., Wen Gong, and Zhan Li. 2008. "Exploring the Digital Divide in Mobile-phone Adoption Levels across Countries: Do Population Socioeconomic Traits Operate in the Same Manner as Their Individual-level Demographic Counterparts?" *Journal of Macromarketing* 28: 397–412.

Yu, Liangzhi. 2006. "Understanding Information Inequality: Making Sense of the Literature of the Information and Digital Divides." *Journal of Librarianship and Information Science* 38(4): 229–252.

CHAPTER 5

Transactional and Transformational Leaders: Their Influence on Technology

Transformational leaders elevate motivation, morale, and integrity of their followers while transactional leaders provide the immediate self-interests of their followers. Northouse (2018) asserts that considerable empirical research supports the utility of distinguishing between the two leadership types. He further articulated that the need for transformational leaders has increased dramatically as a result of changes in the market and labor force and less emphasis is placed on transactional leadership if firms intended to remain effective. Leaders were encouraged to empower their followers by developing them into high involvement individuals and teams focused on quality, service, cost-effectiveness, and quantity of output.

This work recognizes that initiatives such as the adoption of innovations like technology are contingent on elements that have been identified as leadership needs in the mainstream. For example, the successful introduction, implementation, and prolonged use of technological tools require the creation of "high-involvement individuals" and cost-effectiveness. It must be explained here that the transformational leader is seen as one that can move/uplift beyond immediate self-interests through *idealized influence*, *inspiration*, *intellectual stimulation*, or *individualized consideration*. Given that the focus of this work is technology adoption, intellectual stimulation is determined as the most critical construct from the four outlined above. This is most discernible when the leader helps followers with becoming more innovative and creative. *Idealized influence*, *inspiration*, and *individualized consideration*

A. Spencer, *Technology Adoption in the Caribbean Tourism Industry*,
https://doi.org/10.1007/978-3-030-61584-0_5

(one to one attention to followers), both address the kinship and rapport that followers (employees) assign to leaders. They are also responsible for followers choosing to align themselves to the goals of the leader, except for *intellectual stimulation* which addresses a cognitive and behavioral change in the follower. It is reasoned that for the employees to accept, understand, and participate in innovative changes in the organization, leaders must be able to inspire creative behaviors and thoughts.

Although a dominant construct which is applicable to this research emerged, all the categories were explored in the second phase to facilitate bold statements as to whether the leaders investigated fit exactly into any of the two leadership typologies. This concept deals with issues of trust, respect, and admiration for the leader. This research finds that there is an indirect relationship among technology adoption level within firms this is indicative of whether employees are likely to themselves to the leader's vision. This may be associated with new innovations; yet it is a more general issue of kinship and whether employees are loyal to their leader which will inevitably affect how quickly they (employees) buy into the vision for a new tool or process. Idealized influence like charisma is assigned to leaders who followers have great admiration for and the presence of this characteristic elicit follower's inclination to accept and join the leaders desired change processed. The leaders under investigation did not perceive their employee's admiration as being critical to the success of their firms. The general view was that employees were being paid to do their jobs and therefore it was irrelevant whether they held them in high regard or not, if the work was done. Some respondents stated:

> I am not here to be their friend. This is a business and I am more interested in employees doing what they are paid to do. I just want a profitable business and not one in which I am a role model.
>
> I can understand that everyone has their own issues but I cannot be expected to be a counselor and mentor when I have to focus on making sure that we are still in business from day to day.

Short-term orientation of the respondents emerged as a recurrent feature of the discussion. Except for two of the respondents, the others were not particularly interested in the long-term acceptance of the vision but affix their focus on the daily transactions and activities which employees performed. This is characteristic of transactional leaders who focus on the immediate tasks and attaining short-term goals. Conversely, while respondents felt that employee's admiration was not important their respect was paramount. This facilitates and ensures adherence to immediate processes and the following of directives regarding what to do in the firms. An illustrative statement from a respondent was "while you don't have to like or admire me you must show me respect." Therefore, the display of disrespect results in the employee immediate termination. Engendering followers to align themselves to the leader's vision for the firm through coercive as well as admiration for the leader and leadership style may be a more sustainable approach. This approach is particularly essential where major change will occur in the firm. For example, if the firm was to introduce a totally new platform for marketing and sales which makes traditional approaches obsolete it would be easier where the followers believe in the goals and vision of the leader.

The issue of trust was conceptualized by the leaders as belief by their employees that they are truthful and reliable. Importantly, they wanted their employees to believe that they "say what they mean." This encompasses providing truthful accounts of an event or incident and delivering on promises. Interestingly, the utilization of rewards and incentives for the performance of specific tasks has been affiliated with transactional leadership (Pawar 2003). Conversely, transformational leadership appeals to employee's intrinsic motivation which allows them to exceed minimum expectations and see the vision comprehensively. Owner-managers do not ascribe equal attention to the three constructs required to create idealized influence in their firm. While respect and trust are revered as critical elements of their leadership style, very little regard is given to admiration as illustrated in the quotations above. This arises from a limited view and understanding of the definition of admiration. Many of the leaders saw it as whether their employees "like me as a person." However, admiration goes beyond this, it encompasses the ability of a leader to set high expectations and standards and lead by example in accomplishing these standards. It's beyond liking a person but instead being impressed by their actions and abilities. The data analysis reveals that most of the leader's rank unimpressively with the idealized influence which facilitates an environment where new initiatives obtain greater acceptance and participation of the followers, who align themselves to a charismatic leader.

The construct of ***inspiration*** or ***inspirational motivation*** relates to the leader's ability to convey their expectations to employees in such a manner that it galvanizes their understanding and actions to attain the stipulated expectations. The application of this construct in this research was done through an exploration of the general methods to communicating and achieving expectations, specifically how these relate to technological adoption and diffusion in these firms.

The owner-managers conduct orientation sessions or one-on-one meetings to outline and clarify their expectations of employees. However, they expressed that there are times when tasks have been performed and there was insufficient time to ensure that the employees obtained a complete understanding. An assessment subsequent to the completion of the tasks. Often this results in time wasting, trial and error and employee frustration about the processes. A respondent articulated "sometimes because of the nature of our type of business we do not have a lot of time to spend doing non-transaction type activities." This indicates that very little time is allocated to the preparation of new initiatives and in attempting to understand the responses of employees to the introduction of new technologies.

The scenario represents one in which the leaders themselves are not excited about new technologies and therefore owner-managers do not expectations of their use is not effectively communicated. "Most times I cannot genuinely promote these new things because I am not impressed by them." As it relates to communicating expectations, in addition to orientation and training sessions, leaders who introduced websites to their firms were asked to describe the processes undertaken. The primary approach taken was one that involved external website developers conducting an orientation session after which employee would be encouraged by owners to become familiar on their personal time. In some instances, incentives were offered to employees who utilized the websites to increase sales. However, in a commission-driven environment, the lure of incentives is perhaps unattractive to agents considering that these websites were not equipped to facilitate online sales. The sad reality is that owner-managers insufficiently stimulate employees to utilize the technologies because they did not attach enough significance to the adoption process. Consequently, agents decided to focus only on what affected their earnings. Rogers (2003) posits that types of innovation decisions are collective, authority-driven or optional. The utility of this generalization for an example is very useful as the assessment of these firms reveals that comprehensive adoption for the social system is rarely achieved by optional innovation decisions. Collective innovation decisions are predominantly leadership driven although there is agreement by the team. A charismatic leader (Hitt et al. 2001) or an opinion leader (Rogers 2003) usually facilitates this group consensus.

As a corollary response, the agents chose to focus only on what affected their salaries. Rogers (2003) denotes that types of innovation decisions are optional, collective, or authority-driven. This generalization for example is quite useful as the observation of these service firms reveals that *optional innovation decisions* very rarely result in comprehensive adoption for the social system. *Collective innovation decisions* usually gain more traction but are largely leadership driven even though there is consensus by the team. This consensus usually stems from a charismatic leader (Hitt et al. 2001) or an opinion leader (Rogers 2003).

The reality is that while *authority innovation decisions* may yield compliance, those with power may also be opposed to the diffusion of a particular innovation. This again reinforces the importance of leadership in the innovation diffusion discourse. It is then fair to say that collective and authority innovation-decisions are more common in many organizations. Moreover, these processes are leader-initiated, which calls for an emphasis on the orientation of leaders and decision-makers. The disinterest which was displayed by leaders led to a general lack of enthusiasm for the employees and this explains why the majority of firms which had started websites reported that these were now inactive sites. Even so, some respondents stated that they get employees to understand their strategic vision by placing a vision statement in a visible position of the office. However, they did not follow-up to see if employees in fact understood what these statements were saying or if they shared in this vision. Generally speaking, vision statements are most times broad and vague statements. They do not address specific initiatives such as the use of a new technological tool to better serve markets. In order for complete diffusion and adoption of technologies to take place therefore, there needs to be greater focus on communicating expectations, training employees on how to use technology effectively, and monitoring and evaluating their use.

The firms in the study are small firms with the majority having less than ten (10) employees. This makes conditions favorable for individualized consideration or one-on-one attention to be given to each employee. According to Bass et al. (2003), this type of attention shows that the manager cares about the needs of the employee. For Mandell and Pherwani (2003), this is referred to as emotional intelligence. The authors explain that employees are likely to be more loyal to a manager if they get a sense that the manager sincerely cares about them and wants them to succeed. While this does not automatically translate into adoption behavior, it sets a foundation upon which leaders may seek to engage employees in processes which may be new or innovative. This statement assumes that other conditions such as training, coaching, and leading by example are also in place. Owner-managers maintained that they were too integrally involved in operations that they could not give individual attention. Individualized consideration takes a considerable amount of time and energy and involves knowing each employee, their abilities, their personal and professional goals, and what motivates them to perform at an optimal level. The respondents, while recognizing possible long-term benefits, acknowledged that they were only able to focus their efforts on activities that would bring relatively immediate benefits to the firm. "As I said before I have to concentrate my efforts on running the day to day business." Some managers also felt that it was a considerable waste of time to intimately know each employee when the attrition rate was so high and employees would likely resign or be fired soon.

It is evident that the direction of the causal relationship may be blurred in the minds of the respondents. It follows logically that the lack of connection with employees is a causal factor in the high attrition rates. Thus, paying closer attention to the relationship between employer and employee may create longer lasting relationships. Instead of viewing the attrition as the effect, the respondents highlight it as the cause for their reluctance to engage with their employees since relationships are typically short-lived. In an environment where leaders and followers are disconnected, the respondents pointed out that their approach to getting employees to perform well at their tasks was through authoritative coercion or incentives. This is indicated by one of the owner-managers who stated "I think their salary should be their motivation and if they do not meet expectations then we cannot work together." Notably, motivation drivers are being applied without an understanding of what drives each employee. Without truly knowing the employee, the wrong motivating factor is likely to be ineffective. Thus, employees may complete tasks to the bare minimum.

It is particularly striking that these approaches to encourage performance are taken as normal everyday activities for two reasons. One is that the use of incentives to encourage employees to carry out their normal duties is characteristic of a transactional leader, who simply focuses on a task at a specific point in time and does not act as a visionary. Hence, if employees are being paid to carry out particular activities and then each task has to have an incentive appended, then the firm will lose considerable revenues. Similarly, employees are likely to perform below required standards or not at all if no incentives are offered. Additionally, considering that owner-managers have difficulties encouraging good performance for regular, traditional duties, it is safe to say that an even greater challenge will emerge where new innovative ideas are introduced to the firm. This was demonstrated by the failed attempts at developing and using websites in firms. Perhaps the bigger issue however, is that in the future firms may not readily transition from one level to the next on the adoption hierarchy. Each stage of adoption is likely to yield greater resistance as it usually means that there will be a greater deviation from the norm. For example, a typical firm in this sample that only uses the internet for emailing, marketing, and promoting information to clients is likely to experience resistance at the use of websites for online marketing and sales, and even greater resistance to interactive social media marketing. This reality acknowledges that more creative ways of encouraging buy-in are needed. From a leadership perspective, there must be individualized consideration to bridge the distance between the leader and the followers. Through individualized consideration, catalytic changes may be achieved, and performance levels may increase to maximize the overall potential of the firm, the leader, and the employees.

The existing conditions of manager–employee relationship are symptomatic of transactional leadership at play. This work takes a hierarchical approach with both transactional and transformational leadership rather than accepting the view that these leadership styles are at extreme ends of a continuum. This means that transactional leadership traits provide important preconditions for building transformational leadership skills. This distinction is important because the dominant transactional traits which now exist in these firms, such as meeting daily targets and controlling the work environment do not run counter to the transformational paradigm but rather provide a basis upon which this type of leadership may take the firm from one level of technology adoption to the next. Having made this distinction, it is critical to point out that these transactional leadership traits are simply not sufficient to encourage greater participation in innovative activities as they are primarily focused on maintaining the status quo within organizations. In using this research to offer some prescriptive advice, employees must become more involved.

By doing this, the managers can signal that they want change to occur. In some cases owner-managers were not embarking on a path to change. Instead, they were trying to survive using antiquated approaches in the hope that the industry and markets will revert to what they used to be. In these cases there was no attempt at intellectual stimulation for new and innovative ideas from within the organizations.

Intellectual Stimulation

Based on the Multifactor Leadership Questionnaire, the construct of intellectual stimulation most applies to this research. This is most salient when the leader facilitates innovation and creativity among his/her employees. It was important to explore the remaining three constructs to clearly identify if these leaders would fit into any of the two leadership categories based on all four constructs. While three of the constructs demonstrate the likelihood of followers aligning themselves with the goals of the leader, only ***intellectual stimulation*** addresses a cognitive and behavioral change in the follower. This reinforces the fact that for employees to accept, understand, and participate in innovative changes in the organization, the leader must be able to stimulate creativity. As it stands, firms are grounded in traditional practices which are antithetical to productivity. It is gathered that performance will improve with new ideas. For example, one area which needs to become more dynamic is the use of a singular static platform for reaching customers. A top-down approach is needed to move the firm toward ideas surrounding online platforms and internet technologies that can improve their reach and appeal.

If the owner-managers themselves lack innovative ideas, their duty to their employees is to create an atmosphere where creative thought processes are fostered and allowed to flourish. Additionally a lack of ideas from the managers does not mean they cannot influence creativity and innovative behaviors through their leadership style and initiatives that facilitate cutting-edge ideas. When responding to the questions about how employees are encouraged to generate new ideas or come up with new approaches to job tasks, the managers overwhelmingly stated that innovation and creativity were not encouraged. It was articulated that ideas only came from the top of the organization. Based on the culture of the firms, deviation from standard operating procedures was not encouraged nor wholly tolerated. Moreover, organizational culture did not readily welcome suggestions of how a task or process could be improved. Respondents stated:

> While getting ideas from employees may be a good thing, it is more important to me that they are not distracted from doing what they need to do which would be sales. In any case if they suggest something based on their like or dislike for technological innovations that would not be a priority issue.

> Maybe if someone has brilliant new ideas for introducing more innovative things then they are probably over-qualified for this job and would be a better fit somewhere else. I just want them to get on with the business of meeting customers' needs.

Innovative solutions to problems in the firms were either non-existent or very minimal due to the dependence on owner-managers, who were, for the most part, traditionalist thinkers. Therefore, it was very unlikely that they would create contemporary, cutting-edge solutions. A more open system of idea-sharing would allow for a greater alternatives from which to choose and would also empower employees to feel like they are making a valuable contribution to the development of the firm. If employees feel more included, for instance, if an innovative strategy was generated by a peer, it is likely to make them more willing to become innovative. With the exception of two respondents, it is fair to say that according to the major leadership constructs which were employed, owner-managers ranked low in all the areas and can be generally categorized as transactional leaders who simply seek to monitor and control. The interviews about leadership and technology adoption also revealed that two generic categories are not sufficient to classify leaders in the innovation and adoption of technology discourse.

While the constructs from the Multifactor Leadership Questionnaire were useful in identifying general transformational or transactional leadership traits, it was found that other measurement instruments were similarly useful.

Singh and Krishnan (2007) attempted to develop and validate a new scale on transformational leadership. From their work, it was confirmed that adaptability is critical. The authors used the category of "Conviction in Self" to emphasize that pre-planning and preparation in anticipation for the worst is necessary and stems from the ability to adapt and create strategies that are flexible. Respondents stated that they did not necessarily have a system for determining the effects of their current decisions in the medium to long term. A recurring and important theme was intuition. For the owner-managers, "gut-feeling" was a main driver of how they made decisions considering how the industry operates. A representative response from one respondent highlighted that "I have been in the business for a long time and most times I just know if something feels right. If it feels right and I can afford it then maybe." Although it was not surprising that firms were reactive in nature, this response signals a reluctance to think strategically and plan contingencies. Firms typically dealt with a crisis when it happened. This reactive approach saw a massive decline in the number of businesses as well as in the profit margins of the remaining ones when their major source of income was cut by airlines.

As revealed, contingency planning is lacking in these firms and their reaction is sometimes slow when significant change takes place. One of the changes to which they have reacted fairly slowly is the technology revolution and more specifically the internet explosion. Accordingly, when respondents were asked how they identify when old approaches no longer work and new ones are needed, they admitted that oftentimes it was too late when the problem was identified. Once again the lack of a clear system for evaluating the needs of the firm makes it particularly challenging for the formulation of agile strategies, which allow for fundamental change to take place. Technology is universally recognized as a means of survival for firms because it provides solutions to problems. It may be argued that owner-managers cast doubt on some types of technology for their firms, and therefore make the choice not to expand into those platforms. Such a determination would need to be made based on solid data from needs assessment exercises, rather than what currently exists where it depends on an owner-manager's personal preferences. The evidence resoundingly points to a need for something new to happen in light of previously mentioned conditions under which these firms operate.

These agencies, which have seen a decrease of more than 50% in the number of companies in operation over the last ten years, typically cater to the outbound market. In retrospect, it can be argued that their ability to leverage their importance may derive from an improved capacity to meet the need for an online presence in the country. Furthermore, a greater contribution to the island's tourism sector may present a strong case for governmental collaboration and support. At present, these agencies receive little attention from statutory bodies. Particularly, in a volatile marketplace, there is a greater need for the owners of these firms to become visionaries rather than operate as mere transactional "caretakers" of what appears to be static procedures.

Leadership and Strategy

There has been an abundance of research which connects strategy to organizational success. For example in his seminal work Porter (1985, 2001) points to five competitive forces, which have been extensively used by researchers (David 2007; Kim et al. 2004; Stonehouse and Snowdon 2007; Poon 1993; Wynne et al. 2001) over the last three decades. Much of the research has indicated that the strategies that firms employ are influenced by the competition as well as the markets they exist in. However, there needs to be more emphasis on how leaders respond to intelligence about these two factors. While these are useful arguments, they are both output-driven discussions with very little emphasis on inputs such as ownership, leadership, and resources. Therefore, it is critical that this work fills the gap in discussing input considerations. There were respondents who recognized that leadership response to distinct market trends varied, which makes an investigation of leadership responses to this kind of information very important. Why was there acceptance among some owner-managers that local markets were not ready for online platforms? Why were these firms resistant to the idea of diversifying distribution? In comparison to the others with the same information, why were some these firms more proactive in anticipation of market changes?

Of particular interest was the fact that most managers felt that there is a high level of incompatibility with having a high-tech strategy coupled with a hi-touch one (intense personal interaction) despite compelling arguments by theorists that a multi-platform approach to distribution will yield sound business models. It seems, the owner-managers in the sample were more concerned with a perceived psychological risk of becoming too dependent on the internet (Eastlick and Lotz 1999). They feared losing customer interaction. As a respondent stated: "Too much technology can take away from the personal touch that we normally provide. I would hate for my company to become so reliant on these machines that we forget the importance of people in all of this."

It is clear therefore, that if these owner-managers are to engage in greater internet technology adoption, they will need some form of assurance that online and offline channels of distribution can "peacefully co-exist." Since these owner-managers highlight that they make decisions unilaterally, perhaps there would need to be a change in the leadership of the respective organizations. Perhaps, change could also come from a shift in leadership style which may be fueled by education or greater exposure. This is discussed later in this chapter.

A recurring theme in the interviews was that firms were limited by insufficient resources. The actions of the leader is therefore important in terms of decision-making about resources are allocated (Kraajenbrink et al. 2010). The research findings are complementary to the work done by Kraajenbrink et al. (2010). This is even more salient where firms have significantly limited resources and tough decisions must be made about priority spending in order to efficiently use these resources. The resource limitations which were highlighted by owner-managers focused primarily on human and financial resources. In most cases it was felt that there was diminutive financial resources to invest in a hi-tech strategy. In other cases however, managers felt that where money was available they did not have capable human resources to consistently maintain technological innovations such as websites. This was also evidenced where firms in the minority which had websites stated that these were inactive. One respondent points out that: "Even if I wanted to introduce these high-tech approaches the level of capabilities of my employees would not be able to support these activities. The entire thing would be too much of a big investment." The respondents highlighted that with the limited financial and human resources they were more inclined to use resources in areas that would assist in the day-to-day operations rather than on strategic initiatives which they were not sure would yield significant returns. Technological innovations were seen as tantamount to high-risk investments which were more geared toward the long-term and many respondents argued that it was not considered a priority area since it did not address the immediate needs of the markets being served. For that reason, it is reasoned that owner-managers take a transactional approach to leadership, which is geared toward enforcing procedures and control daily activities.

The situation for the firms under investigation is also a unique one in that the respondents are owner-managers who have a stake in the immediate and long-term benefits of the operations. This work posits that owner-managers in a stand-alone environment are inherently transactional leaders based on the challenges of the industry coupled with their desire to realize immediate returns on investment. Their desires are primarily motivated by the need to maintain their families. This argument extends the argument that Covin and Slevin (1988) make that small business managers prefer to be decidedly risk averse, non-innovative, and passive or reactive. Carland et al. (2007) state that they operate their business as an extension of their individual personality and immediate needs. Their demonstrated reluctance to adopt internet technologies therefore, is evidenced by their refusal to engage in inbound travel owing to a lack of commission now, although this may yield long-term benefits.

The exceptions to this norm were companies that had numerous branches and were established brands or companies which were affiliated with other business in other industries. Overall, prioritization for resource allocation is influenced by owner-manager's personal needs, preferences, and short-term fatalistic goals, which do not consider a strategic vision for these firms. Hence, it is important to create a profile of leaders which identifies characteristics that can enable movement to higher levels of technology adoption for sales and marketing in particular. This can offer much normative utility in the travel industry given that it is an information intensive industry. As far as it relates to owner-manager, small firm, developing country situations where decision-making is so heavily dependent on the likes and dislikes of the individual and ultimately the leadership style employed, this profile can provide a basis of understanding and can therefore lead to solutions. The following section elaborates on the characteristics identified.

Transactional vs. Transformational: Leadership Characteristics for Technology Adoption

Leadership emerged as the most significant factor playing a dominant role in technology adoption for sales and marketing in owner-managed small travel firms. While it may be posited that leadership is influential in various types of organizations,[1] the main characteristics which make this demonstrably more resonant in this study are size and owner-management. Smaller firms may be characterized by a more simplified hierarchy and less bureaucracy in decision-making compared to larger firms[2] which invariably places more control in the hand of a single leadership figure. The leader's decision-making power is also magnified when owners themselves are managers of the firms daily operations. Carland et al. (2007) work

[1] Hitt et al. (2001).

[2] Brown et al. (2007).

noted that leaders of small owner-managed travel firms operate their business as an extension of their personality and immediate needs. Therefore, decisions which affect the firm behavior reflect the leader's preferences and personal circumstances.

This body of work does not attribute marketing and sales technology adoption such as the internet exclusively to leadership, as it recognizes the influence of other elements such as culture, resources, and strategy in the preceding chapters. However, the previously mentioned factors of size and owner-manager influence, make the leadership factor the most significant and powerful driver of technology adoption in the firms studied.

The evaluation of the findings revealed several key variables such as education, previous work experience, technology experience, risk aversion, family composition, and intellectual stimulation. These contributed to the conceptualization of leadership groups termed as resistors, caretakers, stabilizers, reactors, and transformers which is directly applicable to the technology adoption discourse. These typologies are created from various combinations and degrees of the above-mentioned variables. The following chapter will discuss each variable and typology in detail, but they will be alluded to here. Prior to this a comprehensive outline of all the relationships explored will be presented in order to demonstrate the emergence of these variables as critical ones in explaining each new leadership typology.

Leadership Background

The background of owner-managers interviewed revealed remarkable differences especially regarding their education. The research found education level among the thirty-one (31) owner-managers to be low.

As seen in Fig. 5.1, more than half, 55% of owner-managers surveyed did not possess a university degree. Generally, certificates or diplomas after secondary education were the highest level of qualification; these were often issued by a professional travel bodies instead of established universities. The explanatory potential of education on its own was weak. Therefore, this was further investigated using a cross-tabulation of the relationship between owner-managers' qualifications and their perceptions of internet use in sales and marketing. While the relationship explored was not statistically significant, it is useful as it identifies relationships for further qualitative exploration and analysis.

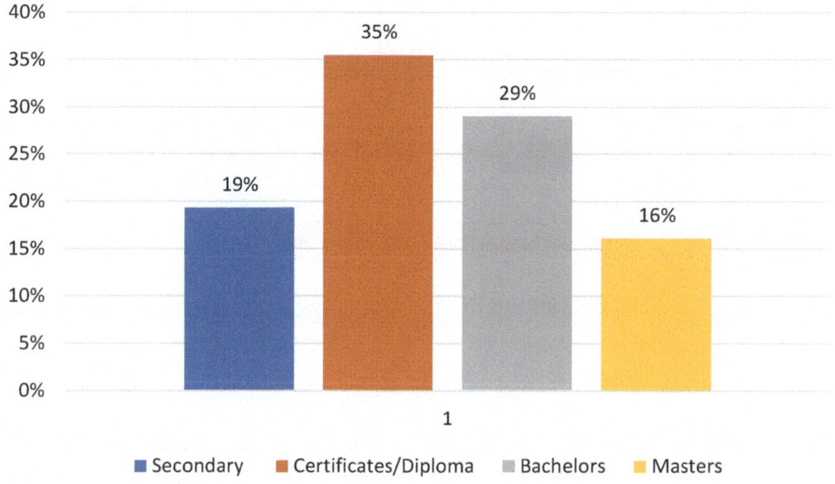

Fig. 5.1 Formal education level (*Source* Author's creation)

Higher levels of education had a positive association with perceptions of the importance in marketing and sales. Figure 5.2 demonstrates that where the highest level of education completed was secondary school, that group of respondents stated that the internet was unimportant or very unimportant. The perception of importance is directly proportionate to education level—as perception increases, educational level also increases. Ultimately, the only two groups that recognized the importance of the internet in marketing and sales were comprised of respondents with a university degree. A similar trend was also detected in Fig. 5.3 which shows the cross-tabulations of education level and perceptions of the importance of the online market to the business.

While none of the group of respondents viewed the online market as being very important, the group of masters educated owner-managers was the only one where all of the respondents felt that this market was important.

Unsurprisingly, those leaders with university qualifications were more receptive to use of different types of technology given their increased exposure to technology use and formal discussions of its advantages in a classroom environment.

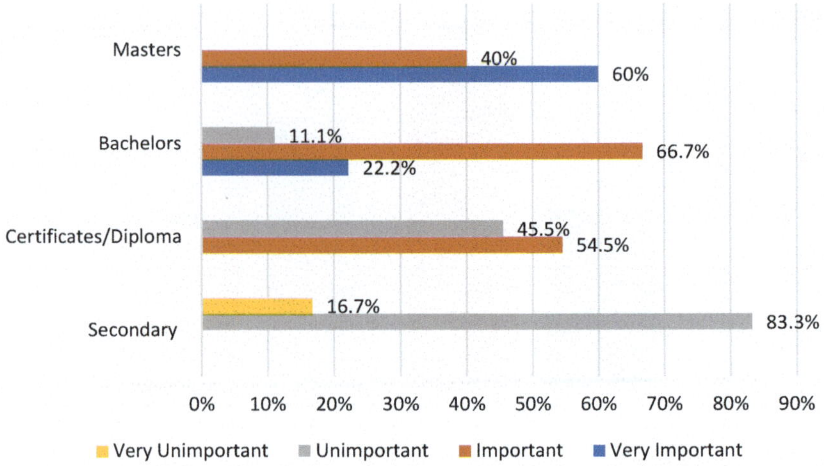

Fig. 5.2 Highest qualification and perceptions of internet importance in sales and marketing (*Source* Author's creation)

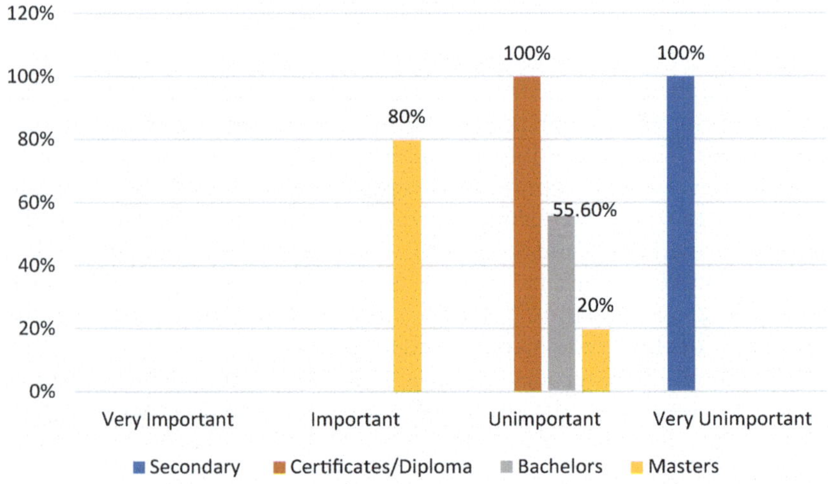

Fig. 5.3 Cross-tabulation: highest qualification and perceptions of online market importance (*Source* Author's creation)

One respondent expressed:

> In a university setting you are forced to use various forms of technology. For instance we had to use the internet to communicate and we even had to write papers on how technology has revolutionized the business world. There are some obvious benefits but there are also some costs.

Further analysis indicated that respondents with a bachelors and masters degree as their highest qualification predominantly had qualification in business and management while other had training in travel and tourism. The most significant finding was the fact that respondents who held a business degree rather than a tourism specific one were more inclined to utilize online platforms for marketing and sales. This may be as a result of greater emphasis on competitive strategies and the development of solid business models in business studies. This finding was gathered from open-ended responses about educational backgrounds which is then compared with technology perceptions. This has provided an exploratory basis for future research on education type and leadership.

In-depth qualitative enquiry indicated respondents who had no university degrees expressed an interest in receiving more formalized training in travel and tourism. Some of the reasons shared included to understand how the industry works as a whole; to understand new and emerging markets; and strategy development. The need for strengthening in these areas is an indication of the weaknesses in the management information intelligence of these firms. Having completed training and earned certificates, these owner-managers felt that this information intelligence gap could be addressed by further training in travel and tourism. The knowledge gap which exists is can be partially attributed to education and training. This may represent a myopic view of how businesses operate and function in a globally competitive environment. The knowledge gap is somewhat related to education and training. Another important variable which emerged in the interviews was experienced. This was explored in terms of proficiency in using different types of information and communication technology. It was felt owner-managers personal experience with using the internet may provide insights into their inclination to introducing technology as a marketing and sales tool in their firms (Fig. 5.4).

The rate of use for owner-managers in the sample was quite low. Internet use among 55% of participants was twice per week while 13%

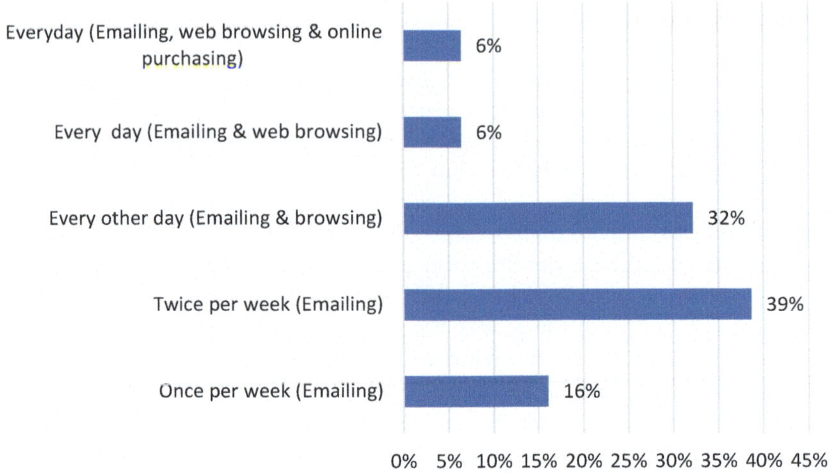

Fig. 5.4 Frequency of internet users (leaders) (*Source* Author's creation)

of the participants used it daily. On its own the rate of use provides little explanatory potential and so the rational related to the use of the internet will add more robustness to the explanations which will emerge from these findings. The participants who used the internet twice per week did so to check their email and organize their inbox while respondents who used the internet every other day checked their emails and surf the net for information. None of these participants ever executed an online financial transaction. Of the four (4) participants who used the internet daily, two (2) of them checked their emails and surf the net while the other two (2) checked their email, surf the net, and used their personal Facebook accounts on blackberries. Of note the more innovative of these two participants executed an online purchase in the past. One owner-managed expressed that he was influenced by his son performing successful online purchases. He noted: "At first I was very wary of doing certain things on the net until I saw my 16 year old son make it look so easy. He was buying music and gadgets online and even though it took me a while I eventually tried it one day and the rest is history."

The owner-managers personal internet usage translated into business practices and firm technology adoption behavior. The only firm which had engaged in online sales was one of two firms which had internet savvy leaders. The other had intentions to develop a better online platform for sales as their current inactive website was an inadequate one. However, these leaders personal Facebook use did not convert into their firm use as they did perceive social media as a business tool. The leaders who engaged in minimal internet use for example, emailing up to two times per week, they experienced low levels of internet involvement and predominantly utilized the web for email activities. Classifications of involvement levels are discussed in detail later in this chapter. The information outlined here was collected in the first phase of the data collection but was supplemented by deeper qualitative inquiries in the second phase uncovered further details about the experience of leaders in using the internet. In recanting their experiences, owner-managers emphasized that they were overwhelmed with information and was uncertain how to navigate through cyberspace. Those with children asserted that their experience of witnessing their children using the web inspired confidence about how to use it. The biggest restricting factors appear to be uncertainty regarding the outcomes of certain activities on the internet as such many respondents had selectively decided to engage in basic emailing and web searches. Most of the leaders of the firms credited this to their family background and exposure which was primarily steeped in traditionalism. Activities such as online purchasing of personal items were viewed as high risk by these owner-managers.

Risk Taking and the Owner-Manager

Risk perception also emerged as an important factor among the respondents. While many respondents view internet buying and selling activities as risky, its unsurprising they perceive investing in platforms for e-commerce transactions for their firms as a relatively high-risk endeavor. To further substantiate this point all the respondents perceive an internet investment as either medium (45.2%) or high risk (54.8%). This is indicative of the owner-managers ambivalence or at the very least uncertainty regarding whether return on such an investment were worth it. This alludes to three types of perceived risk: time risk,[3] psychological risk,[4] and financial risk. All three could be applied in this case as owner-managers

[3] Tan (1999).

[4] Eastlick and Lotz (1999).

were seemingly worried about losing time and money but also psychological risk of increasing their dependence on the internet and loosing the "soft side" of their service delivery. An assessment of how respondents view themselves as a risk taker in a competitive environment is crucial. This finding was determined by asking respondents how they ranked themselves as overall risk takers.

The significant majority of respondents considered themselves to be low-risk takers with only two (2) respondents ranking themselves as high-risk takers in Fig. 5.5. These "high risk-takers" represent those who were mentioned previously as avid internet users (see Table 5.1).

Table 5.1 shows that owner-managers who perceived themselves as high-risk takers or very high-risk takers used the internet extensively for their personal activities including online purchasing. Conversely, those who considered themselves to be very low-risk takers were comparatively low users of the internet. While this study cannot prove causation, this finding indicates that owner-managers with low internet use are the ones less inclined to take the risk of investing in new technologies. This is a serious impediment for internet adoption for marketing and sales in these firms as owner-managers who define themselves as low-risk takers also perceive investment in online platforms as being high risk. So in addition to being risk averse they generally view technological investment as

Fig. 5.5 Owner-manager's risk taking (*Source* Author's creation)

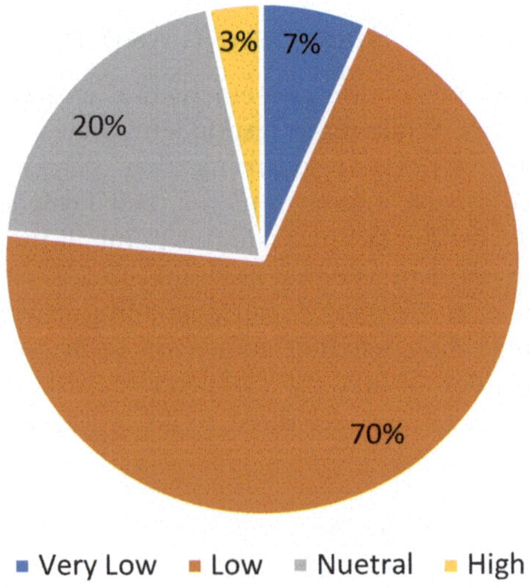

Table 5.1 Cross-tabulation: personal technology use and owner-manager's risk taking

Average rate of usage	*Very low-risk taker (%)*	*Low risk (%)*	*Neutral (%)*	*High-risk taker (%)*	*Very high risk taker (%)*
Once per week (Emailing)	40	60	0	0	0
Twice per week (Emailing)	0	83.3	16.7	0	0
Every other day (Emailing and web browsing)	0	70	30	0	0
Every day (Emailing and web browsing)	0	50	50	0	0
Every day (Emailing, web browsing, and online purchasing)	0	0	0	50	50

Source Author's creation

Table 5.2 Cross-tabulation: highest qualification and internet sales investment risk

Highest qualification	*High risk (%)*	*Medium risk (%)*	*Low risk (%)*
Secondary	100	0	0
Diploma	100	0	0
Bachelor	0	100	0
Masters	0	100	0

Source Author's creation

high risk. A relationship was observed between education levels and risk perceptions.

The findings in Table 5.2 reiterate the point that greater exposure levels through the formal education processes impact perceptions about the potential returns on investment (ROI) from engagement in online selling activities. In addition to technological experience and education, the nature of owner-managed firms typically facilitates greater levels of risk aversion. An additional element, inferred through discussion with the participants is that they are reluctant to take risk with the business because it directly affects their families' survival. This can be attributed to the recurring emphasis these leaders of small, autonomous owner-managed firms place on safeguarding "my business." Some respondents expressed they would be risk takers if they were managers in a business they did not own. Given this declaration it can be argued that these owner-managers are not inherently low-risk takers because of their culture and society but rather act out of protection for a firm in which they have invested heavily. Instead their fear of acting or investing in an idea stems from lack of understanding.

Ironically, many of these protectionist behaviors result in a fixation on short-term benefits. For example one respondent highlighted that "I have to make sure at all times that the business which my family depends on is safe so I cannot make hasty decisions." After discovering these tendencies in phase 1 in the data collection process, it was essential to define leadership classifications and characteristics from the literature which may be applicable to the context of technology adoption. Transactional and transformational leadership were the two foremost types of leadership that emerged. The second phase of the primary data collection focused on outlining how these leadership concepts are applicable to technology adoption practices in small owner-managed firms.

Leadership Characteristics for Technology Adoption (Owner-Managed, Small Firms)

Several variables have been identified in this study as well as in previous research[5] that applies to the major leadership categories. Past research has not really considered leadership characteristics in technology adoption. For example, Peterson et al. (2009) identified some transformational leadership traits that would apply to CEOs who lead high technology start-up firms. However, the authors overlooked broad leadership typologies along a hierarchy of technology adoption. Analyzing any leader this deeply at

[5] Burns (1978), Karnes and Chauvin (1985), Bass et al. (2003), Singh and Krishnan (2007).

each adoption level, will be instructive about what characteristic is needed to matriculate.

In the first instance, leadership emerged as the dominant driver for these owner-managed small firms for a number of reasons. By applying the lens of *Critical Social Science*, a more comprehensive approach could be taken in the research. The critical Social Science perspective allowed for an exploration of a number of micro and macro factors that may affect adoption levels within societies and firms. These included culture, the digital divide, strategy, and resources. National culture emerged as a peripheral contributor since it did not explain differences in the behavior and performance of the owner-managers. The digital divide was also found to be a secondary factor given that the Jamaican society has a high level of mobile penetration and in particular smartphone use with equal access within the society for businesses. While there is still a gap in innovation however, the access gap has been greatly reduced. The previously existing access gap related more to the use of computers. The divide however plays a role at the level of the destination management organization in the context of tourism where there is still a need for a more high-tech approach to meet the information and booking needs of potential high-tech tourists.

Those broader macro factors which did not emerge as significant would have been the most influential in explaining constraints on the leader. The other micro factors of strategy and resources are however determined in large part by decisions made in the firms. In the case of these firms, owner-management and the size of the firm result in simple decision-making processes which are driven by these leaders in the organizations. As it was explained earlier in this chapter, the firms' strategies and the allocation of resources are determined by the priorities of the leadership figures. In light of the pivotal role of leadership, some key characteristics were identified at each stage of technology adoption within these small, owner-managed travel firms.

Education

Education level and type were two significant factors outlined in the study. Those owner-managers who were inclined to adopt based on immediate plans as well as a general intention to explore other commercial uses of the internet, tended to have at least a bachelor's degree. Their openness to a greater use of internet technology in sales and marketing derives from previous exposure to technology in a university setting as well as formal discussions surrounding the technology revolution and its potential benefits. While some of these respondents were not using cutting-edge technology at the moment, they were not resistant to using new technologies in the firm. This presents a particularly fertile condition for the birth of new processes in these firms over which they have control, such as website use for e-commerce and social media for sales and marketing. The type of education also played a role in the openness to technology adoption. Interestingly it was found that those owner-managers who had generic business degrees were much keener on utilizing multiple marketing and distribution platforms than those with specific travel and tourism qualification. From this recognition, it can be interpreted that competitive strategies and the development of sound business models are emphasized in business studies. This then raises questions about the adequacy and appropriateness of tourism degrees. This presents an area for future research.

Previous Work Experience

The respondents' professional backgrounds seemed to differ greatly between likely adopters and unlikely adopters. The two most technologically progressive thinkers in the sample had worked outside of the travel industry extensively. The market leader in the sample had extensive experience in real estate and had built and operated a successful company in that industry for many years. The other respondent who had not yet implemented many of the plans but had intentions to do so, was a bank manager for many years who used retirement funds to start a travel agency. Conversely, the slow or low adopters had spent their entire careers in the travel business. Some started their careers as entrepreneurs while others worked for other travel companies and airlines before starting their own firms. With this understanding it is assumed that a greater exposure to business practices in other thriving industries create transferable approaches to competitiveness. Particularly, the real estate and banking industries had experienced an evolution and restructuring, which may have taught these owner-managers some valid lessons. An important claim in this work is not to say that all owner-managers should immediately employ internet sales and marketing approaches. However, this work advocates that there must at least be a certain level of openness to new ideas which may facilitate their endurance in a time where this is being questioned. Overall, the approach of many struggling agencies in this study seems to be "wait and see."

Technology Experience

Owner-managers who had personal experience with using the internet provided compelling arguments explaining their inclinations toward introducing it as a sales and marketing tool in their firms. The rate of use for owner-managers in the sample was quite low as demonstrated in Fig. 5.4. However, on its own, the rate of use does not explain much. Thus, it is assumed that further exploration of the reasons for which it is used adds more robustness to the discussion. The reasons for personal internet use for leaders were very telling as these translated into business practices and firm adoption behavior. The only firm which had engaged in online sales was one of the two firms which had internet savvy leaders who were actively engaged in online purchasing. The other was in the planning phase for platform diversification. It must be noted that those who had personal Facebook accounts were no more inclined to use social media for the benefit of their firms as it was still being viewed as just a "social" tool. On the other hand, the firms whose leaders engage in minimal internet use for emailing up to two times per week experienced low levels of internet involvement and primarily used the web for email activities.

The personal experience of the group of owner-managers was filled with overwhelming information, which made it particularly challenging for them to navigate through cyberspace in an effective way that could enhance decision-making. This resulted in an overall resistance to a phenomenon which they did not completely understand. The final outcome was that low personal technology experience led to similarly low usage in these firms which were essentially an extension of the owner-managers.

Risk Aversion

The perception of risk was another important factor for the respondents. An overwhelming majority considered themselves to be low-risk takers with only 2 respondents ranking themselves as high-risk takers. These "high risk-takers" represent those who were previously mentioned as avid internet users. This presents a serious challenge for the adoption of the internet for sales and marketing in these firms as owner-managers who describe themselves as low-risk takers also view investments in online platforms as being high risk.

An additional element which has been inferred through discussions with respondents is that they are unwilling to take risks with the business that directly affects the survival of their families. This is a feature of the fact that these leaders are owner-managers of small, autonomous firms. They were more deeply involved from an ownership perspective and pointed out that they were likely to be less risk averse if they were managers in a business which they did not own. It may be argued that they are not inherently low-risk takers due to their culture and society but rather act out of protection for a firm in which directly affected their families.

Family Composition

Family composition emerged from simply asking about the respondent's backgrounds. Inferences revealed that the leaders which were more interested in greater technology use were those with teenage children in their households. They highlighted that the experience of watching their children uses the web inspired confidence about how to use it. However, they were still uncertain about the outcomes of certain activities on the internet. This uncertainty restricted their use. Therefore, many respondents had chosen to only engage in basic emailing and web searches. This prompted further exploration to identify if there were differences with those who seemed less interested in increased adoption. The findings revealed that with the exception of one, all of the other low adopters had older children who were now adults and did not live with them or had small children, who apparently did not influence them in this way. It would therefore appear that such a situational factor was more impactful that national culture since it directed leaders toward technology adoption. All of the aforementioned variables in some way relate to situational considerations that generate differences in each individual. One major difference which has been identified is the level of intellectual stimulation which is provided by leaders of low-adoption firms and high-adoption firms; as defined by those who intend to adopt more cutting-edge tools in their businesses.

Intellectually Stimulating Traits

While all of the respondents ranked low on three of the transformational leadership constructs provided by Bass et al. (2003) in the MLQ, there was a noticeable difference in the *intellectual stimulation* construct for owner-mangers who had a greater inclination toward technology adoption. *Idealized influence*, *inspirational motivation*, and *individualized consideration* are all related to creating an atmosphere of admiration, respect, and kinship to the leader. Although there was no difference between respondents in these areas, the *intellectual stimulation* construct showed that the more innovative-minded owner-managers were different and ranked higher in this area.

This is displayed when the leader helps followers to become more innovative and creative. An exploration of the other three constructs was important to identify if these leaders would fit well with either of these two leadership categories based on all four constructs. While the previous three demonstrate the likelihood of followers aligning themselves to the goals of the leader, the only one which addresses a cognitive and behavioral change in the follower is *intellectual stimulation*. The respondents that had intentions of improving online platforms for sales and marketing created an atmosphere in which employees were free to make suggestions that could be used for innovation. An environment such as this is conducive to the generation of a multiplicity of ideas which may help in the formulation of effective strategies that can boost the firm's performance.

Ideally, owner-managers should endeavor to improve in all of these transformational leadership categories. While improvement is expected, this research explains that along with the other variables identified by this investigation, the *intellectual stimulation* element is sufficient to stimulate higher levels of technology adoption in owner-managed, small travel firms. This finding points to another more prominent issue in leadership and technology adoption research. Now more than ever, more specialized leadership typologies are needed as it relates to technology adoption behaviors. This is primarily because the current classifications of transactional and transformational leadership are too broad and therefore do not apply directly in the innovation diffusion and adoption discourse. The following leadership classifications have been developed by the author to help in identifying the leadership characteristics and needs at various stages of technology adoption for small travel firms. These have been influenced by the transformational leadership constructs and the organizational decision-making literature which describe the decision-maker as creator, actor, or carrier.

The following discussion proposes a synthesis of the findings of this research which allows for the profiling of leaders of owner-managed small firms at each stage of technology adoption. This is based on previously discussed characteristics in this chapter. This research is original in the sense that it derives from the gaps in past studies on technology adoption. However, it is not without its limitations. On its own, the present study does very little to account for small firms in which there is a great level of autonomy such as in the case of owner-managers. The third issue is that although there has been some research which has recognized the influence of leadership traits on small businesses,[6] this research failed to identify leadership characteristics as a driver of each stage of adoption.

The work of Thong and Yap (1995) highlighted three important characteristics of innovativeness, IT attitude, and IT knowledge. This work addressed overall IT adoption, as opposed to a hierarchical approach. On the other hand, Peterson et al. (2009) applied transformational leadership traits to high technology environments and focused on generic traits in start-up firms. This research advances the debate on staged technology adoption and the leadership typologies that influence each stage. Each component of the model being constructed will be explained in detail, a coherent, composite model will be illustrated in Fig. 6.7. As highlighted earlier in this chapter, some key leadership typologies emerged regarding staged technology adoption in small owner-managed firms.

Leadership has been identified as the key determinant of technology adoption for small, owner-managed firms. From a leadership perspective, some key variables have been identified as being influential in technology adoption which is important since previous generic typologies failed to take some of these key variables into account. Additionally, these newly developed typologies combined both leadership and organizational decision-making discourses to identify disaggregated constructs that will be used to inform the model that emerges in Chapter 6. In the chapter that follows, the model will identify the various stages along a technology adoption hierarchy where these new leadership typologies may be applied. Each typology, namely **resistors**, **caretakers**, **stabilizers**,

[6] See for example Thong and Yap (1995), Peterson et al. (2009).

reactors, and **transformers**, will correspond with the varied stages of technology adoption in firms and a detailed discussion of how the model was developed will ensue. The overall conceptual development of the research will be explained prior to this section to give foundation and depth to the research.

References

Bass, Bernard. M., Bruce J. Avolio, Dong I. Jung, and Yair Berson. 2003. "Predicting Unit Performance by Assessing Transformational and Transactional Leadership. *Journal of Applied Psychology*, 88(2): 207–218.

Brown, James, Chris Hendry, and Paul Harborne. 2007. "Developing Radical Technology for Sustainable Energy Markets: The Role of New Small Firms." *International Small Business Journal* 25: 603–625.

Burns, James M. 1978. *Leadership*. New York: Harper and Row.

Carland, J.W., F. Hoy, W.R. Boulton, and J.A.C. Carland. 2007. "Differentiating Entrepreneurs from Small Business Owners: A Conceptualization." In *Entrepreneurship*, 73–81. Berlin and Heidelberg: Springer.

Covin, Jeffery G., and Dennis Slevin. 1988. "The Influence of Organization Structure on the Utility of an Entrepreneurial Top Management Style." *Journal of Management Studies* 25(3): 217–234.

David, Fred. 2007. *Strategic Management: Concepts and Cases*. Upper Saddle River, NJ: Prentice Hall.

Eastlick, Mary Ann, and Sherry Lotz. 1999. "Profiling Potential Adopters and Non-Adopters of an Interactive Electronic Shopping Medium." *International Journal of Retail and Distribution Management* 27(6): 209–223.

Hitt, Michael A., R. Duane Ireland, and Robert E. Hoskisson. 2001. *Strategic Management: Competitiveness and Globalization*. 4th ed. Mason, OH: South Western Publishing.

Karnes, Frances, and Jane Chauvin. 1985. *Leadership Skills Inventory Forms*. Scottsdale: Great Potential Press.

Kim, Eonsoo, Nam Dae-Il, and J.L. Stimpert. 2004. "The Applicability of Porter's Generic Strategies in the Digital Age: Assumptions, Conjectures, and Suggestions." *Journal of Management* 30(5): 569–589.

Kraajenbrink, Jeroen, J.C. Spender, and Aard Groen. 2010. "The Resource-Based View: A Review and Assessment of Its Critiques." *Journal of Management* 36: 349.

Mandell, Barbara, and Shilpa Pherwani. 2003. "Relationship between Emotional Intelligence and Transformational Leadership Style: A Gender Comparison." *Journal of Business and Psychology* 17(3): 387–404.

Northouse, P.G. 2018. *Leadership: Theory and Practice*. Thousand Oaks: Sage.

Pawar, Badrinarayan S. 2003. "Central Conceptual Issues in Transformational Leadership Research." *Leadership and Organization Development Journal* 24(7): 397–406.

Peterson, Suzanne J., Kristin Byron Walumbwa, and Jason Myrowitz. 2009. "CEO Positive Psychological Traits, Transformational Leadership, and Firm Performance in High-Technology Start-up and Established Firms." *Journal of Management* 35: 348–368.

Poon, Auliana. 1993. *Tourism, Technology and Competitive Strategies*. Wallingford: CAB International.

Porter, Michael. 1985. *Competitive Advantage*. New York: The Free Press.

———. 2001. "Bewährte Strategien Werden mit dem Internet Noch Wirksamer." *Harvard Business Manager* 5: 64–81.

Rogers, Everett. 2003. *Diffusion of Innovations*. 5th ed. New York: The Free Press.

Singh, Niti, and R. Venkat Krishnan. 2007. "Transformational Leadership in India: Developing and Validating a New Scale." *International Journal of Cross Cultural Management* 7: 219.

Stonehouse, George, and Brian Snowdon. 2007. "Competitive Advantage Revisited: Michael Porter on Strategy and Competitiveness." *Journal of Management Inquiry* 16(3): 256–273.

Tan, S.J. 1999. Strategies for Reducing Consumers' Risk Aversion in Internet Shopping. *Journal of Consumer Marketing*.

Thong, James Y.L., and C.S. Yap. 1995. "CEO Characteristics, Organisational Characteristics and Information Technology Adoption in Small Business." *Omega—International Journal of Management Science* 23(4): 429–442.

Wynne, Clive, Pierre Berthon, Leyland Pitt, Michael Ewing, and Julie Napoli. 2001. "The Impact of the Internet on the Distribution Value Chain: The Case of the South African Tourism Industry." *International Marketing Review* 18(4): 420–431.

CHAPTER 6

The Applicability of an Innovative Theoretical Model and Its Implications

The conceptual framework presented in Fig. 2.2 has been amended to reflect a difference in the feedback loop for firms. Initially when overall firm adoption decision-making was considered, feedback information was viewed as a function of a system, which was designed to identify firm performance and make decisions in a systematic and consultative way based on strategy. It has been recently identified that in small, owner-managed firms the feedback goes directly to a unilateral decision-maker, whose attitudes and personal beliefs determine technology adoption, as opposed to an objective evaluative system. This emerged from the finding that all decisions about technology use especially for performance, are driven by the specialized preferences of an autonomous owner-manager. This depicts the context more broadly and is further discussed in this chapter, where the development of the model in Fig. 6.7 is addressed.

Theory and Concepts

The nuances of technology adoption have contributed extensively to theoretical developments. For the purpose of this research, the aspect of the theoretical foundation that is used here includes the pre-internet and post-internet phase research. The pre-internet phase saw major contributions from Rogers (1962, 1976, 1983, 1994, 2003) who extended the work done by Ryan and Gross (1943). Rogers' conceptualization

A. Spencer, *Technology Adoption in the Caribbean Tourism Industry*,
https://doi.org/10.1007/978-3-030-61584-0_6

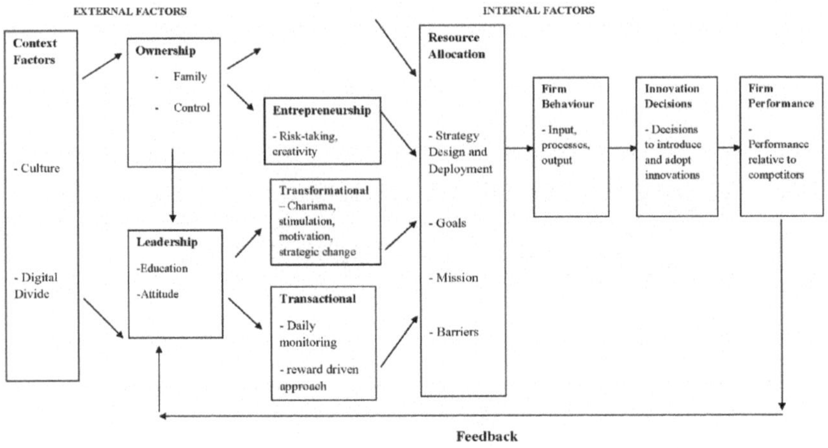

Fig. 6.1 Revised technology decision-making input framework (*Source* Author's creation)

involved four key elements of diffusion as (1) innovation (2) communication channels (3) time, and (4) social system. This provided a platform for the growth of research in the area as it developed factors which were more applicable across fields (Fig. 6.1).

Much of the research indicated that technology adoption was primarily positive.[1] It was only fairly recently that Latzer (2009) identified that some types of new technology may be unsustainable or even disruptive. In the pre-internet phase Davis (1989) became source of authority and was the successor to Rogers. In his technology acceptance model he argues that situational as well as personal influences such as perceived ease of use and perceived usefulness are critical in determining the level of technology acceptance. This claim added the elements of observation and perception to the discussion. However, it was the post-internet era, which sparked most of the technology adoption debate across industries.

Much of this discourse has emphasized a shift from an industrial economy to an information economy, and this may be seen in the development of the literature. Since the internet explosion in 1991, research has been assessing the pervasiveness of the World Wide Web in

[1] See for example Bagozzi (2007).

a variety of industries. There is consensus by many theorists[2] that the rules of competition for established business have been transformed by the internet.

Out of this, many started focusing on reducing information asymmetry, disintermediation, and reintermediation. The last two in particular received significant attention in travel and tourism.[3] This research found that internet transformation was fairly slow in the travel firms that were studied. Later in this chapter, the drivers and barriers to technology adoption are discussed in detail.

In the extant literature, it has been confirmed that various factors drive the adoption of technology. These include national culture,[4] the global digital divide,[5] resources,[6] strategy,[7] and leadership.[8] In both the primary and secondary data collection phases, all of these elements were accounted for and addressed. These factors were conceptualized in two coherent conceptual frameworks (see Figs. 2.1 and 2.2). Figure 2.1 identifies fragmented relationships which exist in the literature which have typically focused on singular relationships. Based on the findings of this research, it is now possible to provide for greater coherence in conceptualizing technology adoption drivers for small, owner-managed travel firms. As a result of this fragmentation in previous studies, it has been fairly difficult to address the gaps in technology adoption in firms. Figure 2.2 illustrates a coherent framework, which places technology adoption decisions within the broader theory of decision-making and highlights relationships which may be explored for future research. This research further emphasizes the role of leadership which is important since this study focuses on owner-managers of small travel firms, who have a great deal of autonomy in decision-making. It must be noted that while Fig. 2.2 influenced the primary data collection process, some of the relationships identified were subsumed in other factors. For instance, ownership was

[2] Rayport (1995), Choi and Stahl (1997), Grieger (2003), Williamson and Scott (1999), Afuah and Tucci (2003), Gentner (2017), Wirtz (2001), Rappa (2002).

[3] See for example Laws (2001), Buhalis and Licata (2002).

[4] Westwood and Low (2003).

[5] Stump et al. (2008), Minghetti and Buhalis (2010).

[6] Brown et al. (2007).

[7] Stonehouse and Snowdon (2007).

[8] Elenkov and Manev (2005), Lynskey (2004), Peterson et al. (2009).

subsumed by leadership. Fundamentally, the research process highlighted that an understanding of adoption drivers in firms lie in the demystification of the decision-making process. The question then emerged: which of these factors influences decision-making most significantly in firms?

The research has led to the creation of leader profiles and ultimately new leadership typologies which have been developed based on the generic transactional and transformational typologies. The new typologies, apply more especially to owner-managers of small firms in information intensive industries. Each typology will be further explained in this chapter. Each typology relates to specific levels of adoption along an adoption hierarchy, also referred to as adoption stages. Stages in this research will be more vertical in nature as they address movement to higher levels of technology adoption.

Adoption Stages

The leadership typologies being advanced relate directly to particular stages of the technology adoption process. There is a considerable body of work about adoption, its stages and particularly for online platforms (see Appendix A). This research will identify stages through which a firm progresses for a particular innovation as well as through different levels of innovation adoption. This approach is different from previous research as it multileveled and it assesses the driver at each level. For example Rogers (1983) highlighted that a firm goes through stages such as: agenda setting, matching, redefining, clarifying, and routinizing for each new innovation. Cooper and Zmud (1990) advocated a six-stage process to include initiation, adoption, adaptation, acceptance, routinization, and infusion. Damanpour (1991) succinctly collapsed these stages into two key areas known as initiation and implementation.

Even the more contemporary research focuses on the stages of a specific innovation adoption. Daniel et al. (2002) focused on e-commerce adoption specifically and created firm clusters in a sequential manner. Cluster one to four involved developers, communicators, web presence, and transactors. This represents an attempt at constructing a vertical process of technology adoption. Moreover, they represent different degrees of e-commerce readiness and adoption through the key variable of technology involvement. Even more recently Aquila-Obra and Padilla-Melendez (2006) have articulated a four cluster approach to internet technology adoption which also emphasizes a sequential process for a single innovation. For a more comprehensive list of adoption models please see Appendix A: Summary of online adoption models.

For the purposes of this research there was a need to conceptualize some technology adoption stages which represented a more holistic demonstration of technology adoption within firms. Unlike previous approaches, this conceptualization does not only consider single innovations and accompanying sequential processes. Rather, it uses a hierarchical approach to highlight different levels of adoption with each level representing a more cutting-edge type of adoption. This was based on the findings of the research for the firms in the sample. While this is not the main focus of this study, it was necessary to develop a hierarchy that corresponds to the new leadership typologies which are ultimately being advanced. Each stage of adoption refers to back-office use and ultimately sales and marketing use at that level. Subsequently a discussion of the leadership profiles will be discussed in relation to these adoption levels. The relationships will be demonstrated more closely in Fig. 6.7.

The Adopters

This research is more concerned with vertical movement along an adoption hierarchy rather than horizontal movement along a sequential process. Therefore, simple phases of initiation and implementation are adopted at each level of adoption. Initiation refers to the point at which firms consider the need for an innovation; they search for information and explore resource allocation needs. Implementation on the other hand, addresses first use of the innovation and the point at which processes evolve to match the new adoption. The broad category of *computer adopters* was used to describe firms at the bottom of the hierarchy that was engaged in simple uses of computer terminals and hardware for back-office accounting functions or front office functions such as sales. Firms which do not use online sales tools other than GDSs are also placed in this category since their adoption was simply based on the free provision of the system by the supplier. It must be noted that computers are also an important pre-condition in firms for other innovation adoptions such as the internet. This was discussed in innovation interdependence in Chapter 2. The researcher acknowledges that all companies had passed through the initiation and implementation phase at this level.

The next group on the adoption hierarchy is termed *internet adopters*. This group represents firms in which the internet is used for emailing and web browsing. A fairly large number of the firms in the sample did not transition beyond this level and only engaged with the internet for client communication and information searches. Most firms already went through implementation of this phase. Conversely, it was observed that some of the firms were still at the point of initiation. *Website adopters* refer to the firms which have created and used company websites for general and marketing information sharing. Websites in this case typically provide static information. While a few firms in the sample had implemented these, they were eventually abandoned and are now inactive. It followed that much of the information became outdated and owner-managers had very little interests in revitalizing these efforts. Outside of those who already implemented the use of these websites, a few other firms were at the point of initiation while the majority of firms were resistant. The next level of the hierarchy is called *e-commerce adopters* and refers to firms which use websites for actual bookings and payment. For this research, this level is considered to be key as it is the level at which there is strident refusal to adopt. The key investigation surrounds why firms are resistant to the notion of online selling. One firm is at the point of implementation while one other firm is at the point of initiation for this level of adoption. An overwhelming majority of firms do not aspire to this level and have no intentions of even exploring the option. This is because there is limited technology experience, limited knowledge of benefits and the perception that the need does not exist. Some responses included:

> I would be wasting my time to try to sell online to my clients because most of them do not like using the internet to do serious business; furthermore they like having that personal contact and they wouldn't trade that for anything.

> I know that there are people who buy online. I have never done so myself and I am kind of wary of doing business with machines. I think my customers also look at it that way and they prefer to talk to someone.

The views of the respondents represent limited approaches to strategy, which do not allow for changes outside of normal activities. The emphasis is on short-termed approaches to keep business alive. This is a direct reflection of the manager who is also the owner of the small business in which there is autonomy to make unilateral decisions. When this is so, what happens is usually protectionist tendencies emerge. These tendencies incline the owner-manager to be more risk averse in the business since it is the business that fulfills the immediate needs of his/her families as well as other interests.

The final level on the hierarchy has been identified as social media adoption in firms. *Social media adopters* would refer to firms in which social media is used for promotion and interaction which results in sales activities. While none of the firms in the sample fall squarely into this category, it appears likely that the firm which is now at the implementation phase of e-commerce adoption will move to the initiation phase of social media adoption. This is based on an expressed intention to explore such an option. It is important to note that social media is also being used on a personal level by this owner-manager. He states:

> I currently use Facebook to connect with friends and I am not using it in the business as yet because I need to find out more about whether customers view it as a serious business tool or just as something to do for fun. I am open to the idea though and will look into it.

Most firms are resistant to moving to any adoption level which is higher than basic internet adoption in the form of emailing and browsing. There are a few firms which have adopted websites for general information and an even fewer number which are exploring online selling. The ultimate aspirational level of social media adoption is still in its infancy and only one firm is even exploring the option. The differences between firms at each level will be highlighted and correlated with leadership profiles.

Contribution to Theories of Staged Technology Adoption

While these categories will assist in the discussion of the hierarchical movement of technology adoption in firms, they are not the essence of what is seminal about this work. The originality of this study emanates from some key gaps in the approaches taken to past studies about technology adoption. The first is that the body of research on stages of adoption takes a sequential continuum approach to understand singular innovations. Secondly, the adoption classifications from previous research tend to make generalizations about firms. While this is useful, it does very little to account for small firms in which there is a great level of autonomy such as in the case of owner-managers. Thirdly, although there has been some research which has recognized the influence of leadership traits on small businesses,[9] the body of research failed to identify leadership characteristics as a driver of each stage of adoption. The work of Thong and Yap (1995) highlighted three important characteristics of innovativeness, IT attitude and IT knowledge that explained overall IT adoption; and not a staged approach at a time when less pervasive business platforms existed. Peterson et al. (2009) in applying transformational leadership traits to high-technology environments focused on generic traits in start-up firms. Through an emphasis on owner-managed small firms, this research works well to advance the debate on hierarchical technology adoption and the leadership typologies that influence each stage. Each component of the model being constructed will be explained in detail. At the end of this discussion a coherent, composite model will be developed. Furthermore, some key leadership typologies also emerged that related directly to technology adoption in small owner-managed firms.

Resistors

Research on the transactional leader has highlighted a short-term orientation in completing tasks at hand by using rewards and incentives to motivate employees to complete tasks. This research however highlighted that among the category of respondents, there are different types of transactional leaders. Clear variables which separate them were also introduced. The **resistor** occupies the lowest level since he/she is least likely to effect change in the firm and is more interested in maintaining traditional approaches. These persons rank themselves as low-risk takers as measured by openness to new ideas from employees, and say that they are willing to do only what is necessary to meet predetermined objectives. Furthermore, their experience with technology is low since they have minimal practice with using computer-related technology outside of infrequent personal emailing; and all of their work experience has been in the travel industry. They tended to have low education levels with the highest formal qualification being at the secondary level. These leaders typically ranked low on the intellectual stimulation scale since they did not encourage creative

[9] See for example Thong and Yap (1995), Peterson et al. (2009).

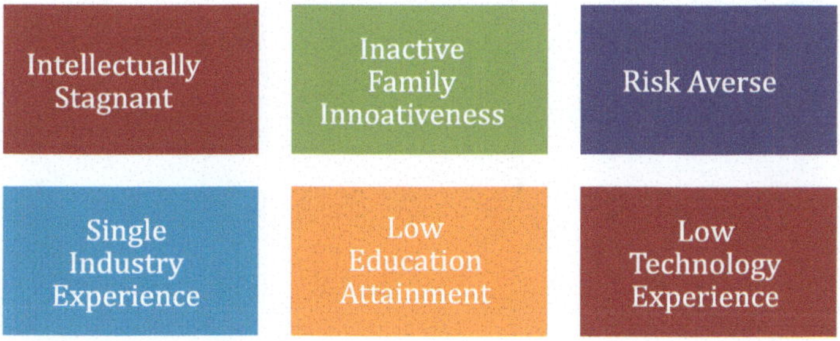

Fig. 6.2 Resistors (*Source* Author's creation)

thinking and innovation among their employees. It is possible that intellectual stimulation in this vein is linked to the make-up of their families. In this sense, where there was no avid teenage internet user whose practice could have influenced perceptions, it was found that leaders were less likely to introduce new technologies in their firms. This trait has been deemed to be the most applicable transformational leadership trait to technology adoption research as it is the only one which appeals to cognitive change and an openness to creativity and new ideas. The researcher coined the term "intellectually stagnating" to signify the converse of intellectually stimulating. Overall, these respondents view internet technology adoption for sales and marketing as an example of a high-risk investment. For these respondents, it meant that there was less proclivity toward adoption. These leaders were mainly found in firms which were categorized as computer adopters with their highest level of interface being the use of Global Distribution Systems (Fig. 6.2).

Caretakers

Caretakers seek to enforce the status quo and control activities to ensure adherence to guidelines. These leaders are transactional and also rank themselves as low-risk takers who will only do what is necessary to follow internal procedures regardless of changes in the external environment. The researcher maintains that this makes them just as intellectually stagnating as the previous group. In fact, they are only different from *resistors* to the extent that they are willing to do things differently, if there is a complete industry change which warrants new procedures. Once new

Fig. 6.3 Caretakers (*Source* Author's creation)

procedures are developed, they will choose to alter any activity within the firm. Caretakers typically had low technology experience like *resistors*. However, they attained a higher level of education than the resistors since they had completed post-secondary studies and earned certificates and diplomas. Certificates were usually awarded by professional travel bodies. The moderate education level gave them exposure to technology benefits although they had limited experience using it. They however had no experience working outside of the travel industry and therefore had a limited view of business strategy. These leaders also ranked low on intellectual stimulation for employees as they did not encourage creative thinking and innovation. Their family composition also tended to be void of an avid teenage internet user whose practices could have influenced perceptions.

Despite changes in the global marketplace, these leaders are not strategic nor are they visionaries. This is because they are insistent on upholding preset standards and procedures. These leaders were found in *internet adopter* firms since they had greater exposure to higher education than the *resistors*. Although they have adopted the internet, this process was slow. Moreover, internet use serves the basic functions of emailing and web browsing (Fig. 6.3).

Stabilizers

This type of leader is still a transactional leader but more closely resembles the *decision maker as actor* in the organizational decision-making literature. This leader is a passive actor who adheres to the firms' operations and tailors his behavior to suit the state of the business at a given time. These leaders consider themselves to be medium risk-takers who will make decisions which may lead to short-term changes to address immediate opportunities and challenges. This may lead to doing something in the firm that is not procedural, however once the incident has passed it becomes business as usual once again. Stabilizers typically had low technology experience and fairly high education levels. Most of them had a bachelor's degree in travel/tourism which seemed to limit their scope. Interestingly, their work experience was similar to *resistors* and *caretakers* in that they did not have any work experience outside of the travel industry. Their family composition had teenagers using the internet which made them more familiar with its use, but they were still no more intellectually stimulating than the previous categories as they usually tried to stabilize situations without employee consultation. For instance, in response to the competition that introduced websites to their firms these leaders who were heads of *website adopter* companies introduced websites also but lacked the proper website design and did not undertake orientation and training. As a result, those websites quickly became inactive and returns on investments were not realized. These leaders had no intention of revitalizing their website efforts and did not intend to move any higher along the adoption hierarchy unless a crisis emerged (Fig. 6.4).

Fig. 6.4 Stabilizers (*Source* Author's creation)

Reactors

This category is representative of a leader who welcomes change and is a medium risk-taker. Although he is not usually resistant to doing things differently, this leader lags behind in the uptake of some new innovations. There is always the intention to improve business practices but these are sometimes late in coming to fruition. This shows the limitations that occur between the initiation and implementation phases, which is usually achieved after much deliberation. Education levels are usually high in this category with a minimum of a bachelor degree with a broader focus than just travel and tourism. Technology experience is moderate (daily business emailing, and web browsing) and previous work experience involves working in other industries apart from travel and tourism. They typically belong to families with avid teenage web users and provide an intellectually stimulating environment for employees but tend to spend a considerable time in the initiation phase of adoption (Fig. 6.5).

This type of leader may move the firm to become e-commerce adopters. However, more data is required particularly since this will involve adopting a technology activity for the firm of which there has been no personal engagement. They differ from the *transformers* in that they are reactive rather than proactive. However when provided with sufficient stimulus they are open to change. Their reactive nature may be changed through increased exposure to technology that might cause them to be less risk averse. It is still maintained here that a greater level of transformational leadership is needed to move the firm to relatively new business platforms such as social media. This will require a proactive visionary approach. Currently, owing to traditionalist perspectives, social media use is virtually nonexistent in the context within which they operate.

Fig. 6.5 Reactors

Transformers

Transformers are considered to be an active change agent with a long-term vision for the firm. They are seen as visionaries and are mostly proactive leaders who are willing to change processes and approaches not only in response to external changes but more importantly, in anticipation of those changes. Transformers are considered to be high-risk takers with high technology experience and a high education level. The high technology experience and high-risk taking are two things that separate this kind of leader from the *reactor*. Moreover, this individual has engaged in personal online buying and is therefore familiar with online transactions. All other variables are similar to the *reactor* such as work experience, family composition and intellectual stimulation. The key difference however, is that these leaders have their own experience with using cutting-edge technology s well as they are able to observe the benefits of technology use by their children. Although these leaders do not use social media in the businesses, they use it personally, thus indicating the reach of these tools (Fig. 6.6).

It must be noted that while transformative leaders are needed to move owner-managed small firms from mere computer adopters to social media adopters—for sales and marketing—the only leader in the sample who is most likely to move to this category has just begun to explore the option and has expressed an intention to adopt. Nevertheless, this firm is still a leader in internet technology adoption for sales and marketing in the industry. This is indicated by the strategic vision and leadership of its top executive and owner. The composite model will now be illustrated and discussed below with an elaboration of the gaps being filled by this research.

Fig. 6.6 Transformers (*Source* Author's creation)

The illustration in Fig. 6.7 is a representation of this work's overall contribution to theory. It is a unique mix of the previously explained leadership typologies and technology adoption hierarchy which emerged from this study. This composite diagram seeks to provide a basis of explanation for different types of leaders who occupy distinct roles based on the stage they are at with technology adoption. Furthermore, it intends to highlight variables that separate each level. The model will assist in identifying potential adoption behavior in small, owner-managed firms in information intensive industries and understanding what factors and/or situations may influence change. While the study was carried out using travel firms, the implications are far more extensive. Having discussed the

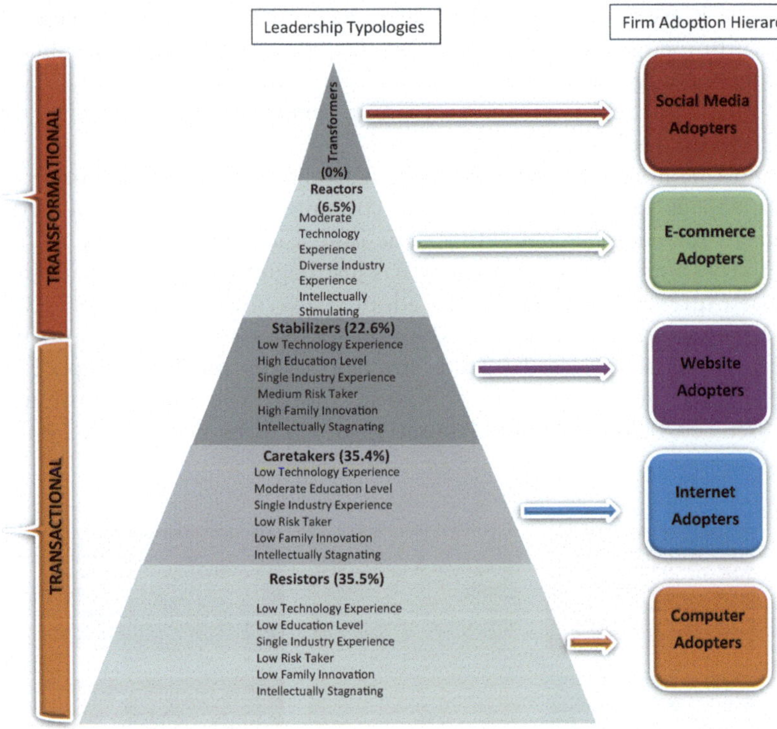

Fig. 6.7 Leadership typologies for staged technology adoption (owner-managed small firms) (*Source* Author's creation)

construction of the model throughout this chapter, it is equally imperative to discuss its strengths and limitations. The subsequent section will address the strengths of this study in relation to theoretical implications and the gaps in the body of literature which it will fill. Inherent limitations of the model will also be presented and contrasted with how some level of delimitation is achieved. The following section will also describe the practical implications of the model. Ultimately the chapter will conclude with a reiteration and emphasis on the work's originality and theoretical contribution.

Strengths of the Model

The model presents a coherent hierarchical adoption approach which identifies the different types of leaders found at each level of adoption. Most importantly, this model is a product of multiple factors that were presented in previous research as potential drivers of technology adoption. By applying a *Critical Social Science* perspective, which called for a more comprehensive assessment of a multiplicity of possibilities, this model could have been developed. As demonstrated in Fig. 6.1, factors such as culture, the digital divide, resources and firm strategy were also examined and secondary and primary data were collected on these factors in relation to technology adoption. The primary data analysis from two phases of collection and analysis revealed that the roles of culture and the digital divide were minimal. In that, there was an indirect correlation between the two factors the individual's perceptions. None of the two was able to explain why similar cultural backgrounds and IT access resulted in different behaviors among the firms.

The resource and firm strategy factors emerged as peripheral drivers as they were directly under the control of the decision-maker in these firms and appeared more as an effect than a cause. The research journey revealed that the critical direct driver of technology adoption in these firms was its leadership. This was primarily so because of the small autonomous nature of the firms in the study. This uncovers another important strength of the project as it is able to fill the gap where owner-managed small firms have been ignored in the technology adoption discourse of past studies. The model that was developed in this research emphasizes that where leaders of small firms are the owners themselves, innovative behavior is significantly challenged. First of all, the small size of the firms usually produces a simple management structure and hierarchy which allows for less bureaucracy and faster decision-making. On the surface, this appears to be positive. However, it also means that decisions are made based on attributes of character. For instance, making decisions hastily to satisfy specialized preferences are likely to become characteristic of the firm's behavior. Seeing that as the owners of small firms, these top executives have an immediate stake in the outcome of business activities, they are more likely to personalize decision-making rather than rely on logic and strategy. From this recognition, it is concluded that the leaders themselves may be initiators or barriers to technology adoption based on their attitude to technology.

Having identified that the leadership element is most critical to technology adoption in owner-managed small firms, the position that this research takes acknowledges that a more detailed approach is to be taken to identify what types of firms were at the various levels of adoption and what type of leader was responsible for the firm's position along the adoption hierarchy. While there has been previous research on stages of adoption, much of this research has been focused on the sequential adoption of specific technological innovations.[10] This research however takes a hierarchical approach. The model it uses therefore is an illustrative of the different technology adoption processes and therefore provides a macro approach to this conceptualization.

The major theoretical development is that the work goes on to develop leadership typologies at each level of adoption, which has never been done before. It must be stated that there were a few odd cases in the interviews which did not match the criteria of a particular typology based on the variables which were used to create the types of leaders. These were however in the minority. Other resonant themes could be identified across interview manuscripts to make these bold conclusions. This takes the research a step further than previous research which identified the importance of leadership in innovation diffusion and adoption.[11] Much of this body of work focused on the initial adoption of innovation in general or on a single innovation. The development of leadership typologies is particularly useful based on the limitations of the transactional and transformational typologies. For example, except for intellectual stimulation, which has

[10] See for example Cooper and Zmud (1990), Damanpour (1991), Daniel et al. (2002), Aquila-Obra and Padilla-Melendez (2006).

[11] See for example Thong and Yap (1995), Peterson et al. (2009).

been used in the development of this model, none of the other constructs identified in transformational leadership research could be directly applied to innovative behavior. Additionally, these typologies, which for the most part were generic, seemed to overlook important variables although they too were applicable to the innovative behavior of leaders. These variables are technology experience and innovative family behavior.

In the case of owner-managed small firms where the leadership element is most critical, it is of paramount importance that an applicable set of classifications which consider more technologically related variables be advanced in order to understand why some firms are at higher levels than others. This was particularly interesting in the travel context, where firms are most at risk of facing exogenous global shocks which could affect their stability and even undermine their relevance. Even with impending danger to the survival of businesses, some firms were still at relatively low levels of adoption along the hierarchy and it was interesting to understand why this was so in such an information intensive industry. Although the study focused on travel firms, the findings may be generalizable. This means, the results could be extended to other firms that also exist in information intensive industries which are small and managed by their owners. It is also possible that lessons can be learnt about leadership drivers and barriers to technology adoption. The model provides useful information for small owner-managed firms in the travel industry as well as those with similar characteristics in other information intensive industries. The leadership typologies presented are instructive to firms that aspire to become more competitive. They highlight key variables which explain individual attitudes and behaviors regarding technology adoption in firms and can empower industry practitioners to claim for the highest level of transformational leadership. Furthermore, they can signal a need to become as proactive as the *transformers*. This will be particularly useful when these firms are anticipating competitive challenges and opportunities. Owner-managers of small travel firms are now able to explore their leadership style and traits through the lens of this study and assess whether their approaches to operations and strategy are directly linked to education, technology experience, work experience, family composition, risk aversion, or level of intellectual stimulation (openness to new ideas from employees) which they provide in their firms. The ability to change one or more of these variables may result in a change in decision-making processes related to technological innovations and more specifically online selling.

Limitations of the Model

The model used in this study makes a significant contribution to the understanding of leadership influence in levels of technology adoption. However, it is not without its limitations. First, the study was conducted with a sample of small owner-managed firms in the travel industry for one country. Additionally in designing the leadership typologies, there were two outliers in the data set for two of the typologies. This limitation acknowledges that transferability extends only to similar companies, similar industries and countries that exist in a similar context, for example other developing countries. In this sense, transferability would not be too difficult since a large number of firms in developing countries are small, owner-managed and operating in information intensive industries. The second more theoretically important issue is overridden by the resonance of the key themes that emerged from each leadership grouping. While there were differences in the variables for one respondent in each of two groups, a pattern in the overall responses was still clearly identified and the robustness of the findings was not compromised.

References

Afuah, A., and C.L. Tucci. 2003. *Internet Business Models and Strategies: Text and Cases* (Vol. 2). New York: McGraw-Hill.

Aquila-Obra, Ana, and Antonio Padilla-Melendez. 2006. "Organizational Factors Affecting Internet Technology Adoption." *Internet Research* 16(1): 94–110.

Bagozzi, R.P. 2007. "The Legacy of the Technology Acceptance Model and a Proposal for a Paradigm Shift." *Journal of the Association for Information Systems* 8(4): 3.

Brown, James, Chris Hendry, and Paul Harborne. 2007. "Developing Radical Technology for Sustainable Energy Markets: The Role of New Small Firms." *International Small Business Journal* 25: 603–625.

Buhalis, Dimitrios, and Maria Cristina Licata. 2002. "The Future eTourism Intermediaries." *Journal of Tourism Management* 23(3): 207–220.

Choi, Soon-Yong, and Dale Stahl. 1997. *The Economics of Electronic Commerce.* Indiana: Macmillan Technical Publishing.

Cooper, Randolph B., and Robert Zmud. 1990. "Information Technology Implementation Research: A Technological Diffusion Approach." *Management Science* 36(2): 123–139.

Damanpour, Fariborz. 1991. "Organizational Innovation: A Meta-Analysis of Effects of Determinants and Moderators." *Academy of Management Journal* 34(3): 555–590.

Daniel, Elizabeth, Hugh Wilson, and Andrew Myers. 2002. "Adoption of E-Commerce by SMEs in the UK: Towards a Stage Model." *International Small Business Journal* 20(3): 253–270.

Davis, Fred D. 1989. "Perceived Ease of Use, and User Acceptance of Information Technology." *MIS Quarterly* 13(3): 319–339.

Elenkov, Detelin, and Ivan Manev. 2005. "Top Management Leadership and Influence on Innovation: The Role of Socio-cultural Context." *Journal of Management* 31(3): 381–402.

Gentner, D., B. Stelzer, and L. Brecht. 2017, June. "Analyzing Customers' Readiness for Digital B2B Business Models." In *ISPIM Innovation Symposium*, 1. The International Society for Professional Innovation Management (ISPIM).

Grieger, M. 2003. "Electronic Marketplaces: A Literature Review and a Call for Supply Chain Management Research." *European Journal of Operational Research* 144(2): 280–294.

Latzer, Michael. 2009. "Information and Communication Technology Innovations: Radical and Disruptive?" *New Media Society* 11(4): 599–619.

Laws, Eric. 2001. "Distribution Channel Analysis for Leisure Travel." In *Tourism Distribution Channels: Practices, Issues and Transformations*, edited by Dimitrios Buhalis, and Eric Laws, 53–72. London: Continuum.

Lynskey, Michael. 2004. "Determinants of Innovative Activity in Japanese Technology-Based Start-up Firms." *International Small Business Journal* 22(2): 159–196.

Minghetti, Valeria, and Dimitrios Buhalis. 2010. "Digital Divide in Tourism." *Journal of Travel Research* 49(3): 267–281.

Peterson, Suzanne J., Kristin Byron Walumbwa, and Jason Myrowitz. 2009. "CEO Positive Psychological Traits, Transformational Leadership, and Firm Performance in High-Technology Start-up and Established Firms." *Journal of Management* 35: 348–368.

Rappa, Michael. 2002. "Business Models on the Web." Accessed February 7, 2010. http://digitalenterprise.org/models/models.html.

Rayport, Jeffrey F. 1995. "Exploiting the Virtual Value Chain." *Harvard Business Review* 73(6): 75–85.

Rogers, Everett. 1962. *Diffusion of Innovations*. New York: The Free Press.

———. 1976. "New Product Adoption and Diffusion." *Journal of Consumer Research* 2: 290.

———. 1983. *Diffusion of Innovations*. New York: The Free Press.

———. 1994. *A History of Communication Study—A Biographical Approach*. New York: The Free Press.

———. 2003. *Diffusion of Innovations*. 5th ed. New York: The Free Press.

Ryan, Bryce, and Neal Gross. 1943. "The Diffusion of Hybrid Seed Corn in Two Iowa Communities." *Rural Sociology* 8: 15–24.

Stonehouse, George, and Brian Snowdon. 2007. "Competitive Advantage Revisited: Michael Porter on Strategy and Competitiveness." *Journal of Management Inquiry* 16(3): 256–273.

Stump, Rodney L., Wen Gong, and Zhan Li. 2008. "Exploring the Digital Divide in Mobile-phone Adoption Levels across Countries: Do Population Socioeconomic Traits Operate in the Same Manner as Their Individual-level Demographic Counterparts?" *Journal of Macromarketing* 28: 397–412.

Thong, James Y.L., and C. S. Yap. 1995. "CEO Characteristics, Organisational Characteristics and Information Technology Adoption in Small Business." *Omega-international Journal of Management Science* 23(4): 429–442.

Westwood, Robert, and David Low. 2003. "The Multicultural Muse Culture, Creativity and Innovation." *International Journal of Cross Cultural Management* 3(2): 235–259.

Williamson, Oliver E., and Scott E. Masten. 1999. *The Economics of Transaction Costs*. Cheltenham: Edward Elgar Publishing.

Wirtz, Bernd W. 2001. *Electronic Business*. Wiesbaden: Gabler Verlag.

CHAPTER 7

Conclusion

The findings of this study are original in their contribution to the literature in a myriad of ways. Primarily, they postulate that research on the adoption of technology in firms requires a holistic approach before key factors are determined. This is unlike the reductionist approach to technology adoption as is expressed in previous studies where positivism is applied. Yet, for a much clearer perspective about the critical emergent factors, the Critical Theory facilitates transparency and inclusiveness, thereby providing for subliminal matters to be addressed before deeper insights can be extracted for clarity. Another significant contribution of the research is the conceptualization of staged adoption through a combination of leadership traits and typologies that dictate the movement of firms along the hierarchy of adoption for single inventions.

While there are other hierarchical representations of technology adoption, consolidating multiple variables into clear typologies is arguably a novel idea, since it details the process of transition from one stage to the next- of technology adoption. Thus, it is fundamentally valuable to previous studies on transformational leadership theory. The results demonstrated that generic classifications of leadership are valuable in identifying leaders as change agents. However, these results did not sufficiently explain how leadership influenced vertical technology adoption in small, owner-managed firms. The model in Fig. 6.7 responds to this gap by further disaggregating generic leadership classifications into smaller categories within the transactional and transformational leadership typologies.

A. Spencer, *Technology Adoption in the Caribbean Tourism Industry*,
https://doi.org/10.1007/978-3-030-61584-0_7

As it is suggested in the discourse, these distinct typologies have generally been more effective primarily in terms of the application to technology adoption.

The study's findings have helped with redefining leadership as it relates to technology adoption by identifying distinct leadership characteristics at each level of adoption.

This research focused primarily on online selling practices which was revealed as the highest level of technology adoption, yet only two agencies were at this level. One of the firms was at the initiation phase of adoption while the other was at the implementation phase. Aspirations for social media adoption were also included since none of the firms in the sample confirmed usage of this tool in sales or marketing. These arbitrary practices reinforced the need for a theory to guide the discourse about the existing relationship at each level of adoption.

Fundamentally, behaviors related to technology adoption in these organizations were interrelated to the decision-making processes about what technologies should be adopted; and the management and implementation of said technology. In this vein, the theory of organizational decision-making was explored to gain a deeper understanding of how long-term decisions which could potentially influence the strategic future of organizations were initially conceptualized in previous research. According to past research, the theory of organizational decision-making unraveled the intricacies of the decision-making process in such a manner whereby the individual became the subject and the focus.

Scholars have categorized decision-makers into three groups; these are actor, creator, and carrier. The actor is reactive and responds to situations while the creator is proactive and innovative-minded. Typically, the carrier is merely a decision-maker who uses previous experiences to gauge the decision-making process; although the same can be said about both the actor and the creator. Based on these groupings, Fig. 2.2 was developed as a framework to demonstrate how decision-making occurred and the implications it had for technology adoption in the firms that were sampled. This framework served a dual purpose in representing an output framework based on the secondary data analysis, and as an input framework for the primary data collection.

The framework indicates that ownership, leadership, and previous experiences affect the overall structure of the firm, including its size, the mode of control and division. These factors also influence creativity and risk-taking which are inherent to entrepreneurship and leadership issues such as stimulation, drive, charisma, and strategic change. Inevitably, strategy formulation, implementation and resource allocation are also affected by these factors. According to Victorino et al. (2006), organizational behavior affects innovation adoption. It is argued in the literature that as input, processes, and output are influenced adoption processes are affected inadvertently.

The view expressed earlier that an understanding of individual decision-making could explain organization decision-making was particularly resounding in this research where the total sample was comprised of small owner-managed firms. Within these firm's decision-making was not constrained by complicated bureaucratic red tapes. However, they were driven by personal choices that characterize the preferences of the owners as shown in their organizational focus. Contrastingly, the organizational decision-making process differs in firms without owner-managers. Notably, key features like risk aversion, technology experience, and family composition are less significant in firms which is not reliant on the direction of a single decision-maker. Thus, in this research, it was deduced that specific attention must be paid to ownership and leadership issues.

The ownership and leadership construct merged due to owner-managers being the primary subject of the research. Majority of their leadership decisions were impacted by their personal investment in the business. In fact, for most owner-managers, their businesses were an extension of themselves. Leadership emerged as the dominant factor in the investigation based on its demonstrated effects on adoption behavior. Therefore, indicating that technology adoption is determined by their overall understanding of ICTs.

The data collected revealed owner-managers perceptions were not in favor of greater utilization of technology in sales and marketing efforts. They viewed technological investments as high-risk which provided low returns. Interestingly, they highlighted the associated benefits of speed, convenience and efficiency with using the internet. Yet, these benefits were overwhelmed by the perception that users may become overly dependent on the internet. Thus, this was recorded as a possible psychological risk of increased technology use. For the most part, owner-managers value the human element in service delivery and feared that the opportunity to interact personally with clients would be lost. A few of the respondents had diverging perceptions which fostered a greater openness and desire to increase adoption. These respondents were of the view that utilization of multiple platforms was the best practice. In this way, a multiplicity of platforms would not only create an avenue for the more technologically savvy customers but also maintain personal contact with those clients who required it. The nuances in these views revealed differences in family background, education, risk aversion, technology experience, and work experience.

Given the critical role of owner-managers' decision-making in their firms, the need emerged to understand the basis of their perceptions and ultimately their approach to their styles of leadership. In the second phase of the primary investigation, the researcher explored transactional and transformational leadership through qualitative enquiry. Critical to the most authoritative work in transformational leadership (Bass et al. 2003), four essential constructs were purported. Intellectual stimulation emerged as the most influential in the discourse of technology adoption. This construct was measured using the responses that were shared about a leader's openness to receiving new ideas from employees. All, but two of the respondents expressed that they did not encourage an open environment where employees could share their ideas. They were more concerned about employees performing their daily functions.

Idealized influence, inspiration and personalize consideration creates an environment in which employees respect, admire, and have kinship toward the leader. Although each respondent echoed the same sentiments in these areas, the intellectual stimulation construct showed that the more innovative-minded owner-managers ranked higher in this area. Innovation and creativity within firms coalesce around a leader who the staff recognizes as intellectually stimulating. This assumption acknowledges that those leaders who welcome their followers' ideas and/or change initiatives are more likely to be successful change agents. By exploring the other three constructs, the researcher was able to determine whether each leader could potentially fit into the two main leadership categories based on all four constructs that were used. Deriving from the three constructs that were investigated, it is evident that followers willingly align themselves to the goals of the leader; although, it is intellectual stimulation which demonstrates both a cognitive and behavioral change in the follower. It is interpreted from these findings that the categories of transactional and transformational leadership were both insufficient to determine leaders' capacity to influence separate levels of technology adoption. This work recognizes the value in approaching leadership in a transformational way, as an attempt to improve and increase technology adoption. At the same time, further disaggregation is required if the emergent leadership typologies are to be clarified. The typologies coalesce around an intellectual stimulation component (openness to employees' ideas) of transformational leadership; and other factors including the level and type of formal education, technology experience, previous work experience, family composition, and risk aversion.

Following assumptions from previous studies about a firm's behavior, this work explored the internal factors of strategy and resources since a firm's resource capabilities (Barney 2001; Kraajenbrink et al. 2010) and its strategy decisions (David 2007; Stonehouse and Snowdon 2007) ultimately affect that firm's behavior. Both secondary and primary data was used to explore these factors and the analysis revealed through a process of distillation that firms with these traits, resources, and strategy were not central factors. However, they are peripheral contributors and are not dismissed. This is because each of these characteristics is used by leaders who seek to influence the process of technology adoption in their respective firms irrespective of whether or not adoption is something they pursue. In a variety of ways, they arise as a consequence of leadership capability and behaviors instead of inherent drivers. They may emerge to be more influential in a different environment but for small owner-managed firms in information intensive industries they are more pawns in technology decisions rather than key determinants.

The investigation of these internal factors through primary data collection revealed that there was no evident strategy which informed technology adoption behavior in local owner-managed travel companies. It can be deduced that these companies engaged in a high-touch strategy (intense personal interaction) but were apprehensive about perceived psychological risk of reliance on technology as purported by Eastlick and Lotz (1999). They held the view that majority of their clientele place a high value on personal interaction, which is influenced by their culture. Furthermore, owner-managers believed that high-tech and high-touch strategy was unsuitable.

From a resource perspective, there are concerns about the constraints that owner-managers experience in terms of human labor and finances. Many owner-managers implied however that even if they had the resources, they may not be inclined to adopt more technologies. The consequence is that the sole decision of resource allocation lies with owner-managers. The decision to spend in areas other than technology adoption usually indicated the leader's interests and preferences. Notwithstanding that majority of the respondents attached importance to technology adoption in their firms, the decision to adopt more technology coalesces around understanding and capacity to use and maintain the technology. The analysis indicated that firm strategy and resources are not enough to explain organizations' adoption behaviors and the differences between them. These factors were especially insignificant in explaining why firms were at different level on the technology adoption hierarchy as the heterogeneity of strategies and resource constraints were not enough to identify why several firms were more technological advance than their counterparts. These internal factors manifested as a consequence of difference in leadership of small owner-managed travel companies rather than as significant contributors when deciding to adopt new technologies.

Furthermore, the context in which industries are embedded may also explain overall industry behavior. For example, while culture and the digital divide may explain behavior, they do not always account for the differences in behavior among firms in the same society. Social interactions, traditions, norms, and values were the constructs assessed. Considering that the firms investigated operate in similar external environments it was concerning that they made different operational and strategic decisions. It was discovered through examination of culture as a factor that firms were thriving in a relationship-oriented society where familiarity and friendships impacted business relations. Added to this, it was found that the national culture had its influence whereby it promoted traditionalism with unwavering adherence to societal norms. Most importantly, cultural factors at the national level could not be used to rationalize the diverging levels of technology adoption among firms since it was experienced by all participants of the sample. Notably, the subcultural influence of the family proves to be a more determining feature of culture.

Family composition influence technology adoption through its impact on the leaders of these companies. While employees' views are also shaped by family, leaders too are heavily influenced by familial background. This relationship was recognized as more important, particularly for small owner-managed firms who make unilateral decisions about innovation. Family composition was also important since it allowed for distinction between high adopters and low adopters of technology.

Owner-managers who shared the household with family and who were engaged in innovative behavior were comparatively more open to new ideas that could be employed for the benefit of their firms. The influence of the family as a subculture explains the large differences but it does not the subtlety between leadership categories. For example while family composition was common among firms, there were both moderate and high adopters of technology. Thus, family structure/the influence of the family must be understood with other factors to explain variances in behavior. The firms in the sample had equal access to technological resources and were equally limited by their understanding of the digital space. Hence, the digital divide appeared to be an insignificant context factor in explaining firm adoption behavior.

Conversely, within the business context of the country a global divide was more evident than a domestic one. In light of this fact, it is not difficult to imagine that these travel firms may reconcile the technological divide (Minghetti and Buhalis 2010) and communication gap (Maurer and Lutz 2011) that occur between destination countries and dominant tourist-generating countries, specifically the United States and the United Kingdom. This may be accomplished by utilizing independent and covert agents such as the internet (Govers et al. 2007).

Contextually, an even amount of exposure to similar factors would indicate that their explanatory power has been reduced and would therefore become more difficult to understand the differences in technology adoption among firms. In this regard, the evenness of influence—of these factors—on firms in the society are instructive only to the extent that they can inform restructuring of the industry. This objective was primarily to determine whether these factors influenced decisions of small owner-managed travel firms. However, these were found to be peripheral factors, meanwhile culture provided only a sub-element of the key determinant leadership.

Essentially, firms are extensions of their individual leaders who own, manage and operate them. Oftentimes, these decisions reflect their individual preferences and experiences. The model as demonstrated in Fig. 6.7 contributes to two main theoretical areas. The first engages the adoption of technology. A key observation is that the literature surveyed on stages of adoption takes a similar approach of sequencing to understanding singular innovations. While this has proved useful thus far, it does not account for multiple levels of adoption (see for example Daniel et al. 2002; Aquila-Obra and Padilla-Melendez 2006). For the purposes of this research, it was necessary to define stages for technology adoption; which embodied a more holistic approach to technology adoption within firms. This approach differs from those of the past since it does not only consider single innovations and the accompanying sequential processes instead the levels of adoption are identified using a hierarchical approach as each level represents a more progressive type of adoption. This was based on the findings of the research. Notwithstanding that this is not the main focus of this study, choosing to develop a hierarchy that corresponds with new leadership typologies adds utility to the discourse here. Each stage of adoption refers to computer adopters, internet adopters, website adopters, e-commerce adopters, and social media adopters which are used mainly for back office, sales, and marketing purposes.

According to the innovation interdependence literature (Moital et al. 2009), each subsequent stage in the hierarchy is treated as a precursor for the other stages that follow. Computer adopters include those firms that employ computer technology for simple tasks such as back-office functions and GDS use. Conversely, Internet adopters include firms that depend on computer technology for emailing and web browsing functionalities. On the other hand, Website adopters engage online platforms for marketing information, inter alia, while e-commerce adopters use digital platforms to transact payments. Lastly, Social media adopters rely on social media for sales and marketing functions. Each of the aforementioned stages of adoption occurs between two main phases—the initiation phase and the implementation phase.

The identification of new leadership typologies is the most profound contribution to the previously outlined hierarchical levels of technology adoption in small owner-managed firms. This correlation between each distinct leadership typology and the separate levels of adoption is illustrated in Fig. 6.7. While there has been some research which has recognized the influence of leadership traits on small business adoption (see for example Thong and Yap 1995; Peterson et al. 2009), leadership characteristics have not been identified in this research as a determinant of adoption at different stages. However, the existing body of research acknowledges that leadership is indispensable to the overall adoption of information technologies in small start-up firms.

The emergent typologies are referred to as resistors, caretakers, stabilizers, reactors and transformers. These typologies indicate a first attempt at disaggregating leadership categories into one model that reflects leadership influence at each stage of a hierarchical technology adoption structure.

Although there were six key variables that influenced leadership, each of them offered varying degrees of influence over each stage of adoption. Education was found to be the most prominent variable in the transition from computer adoption to internet adoption. Based on the post-secondary studies that were undertaken, education level was found to be typically higher with leaders. For website adopters, education, risk-taking, and family background were identified as the most salient variables. When compared to computer and internet adopters, website adopters acquired university degrees, had active technology use at home and were medium risk takers.

In some firms, the transition from website adoption (general information) to e-commerce adoption varies based on technology experience, industry experience and type of education which were generally found to be the dominant variables. Generally, the leader possessed moderate technology experience which involved daily emailing and browsing. The leader also had diverse industry experience and had acquired university education not limited to travel and tourism studies. For those at the e-commerce level who desired to transition to social media adoption, it was found that their technological experience and risk-taking once improved could propel them to the next level. Mostly, ecommerce adopter firms were managed by leaders with moderate technology experience and who were generally low to medium risk-takers. It has been considered that high technology experience can cause leaders to underestimate the risks involved and overestimate their capacity to manage those risks after experiencing the benefits of using online commerce platforms. This could be inferred from the most open-minded respondent who intends to explore social media options while still functioning at the ecommerce level and has high technology experience.

Most importantly this model was developed out of evaluation and comparison of factors that have been represented in previous research as basic drivers of technology adoption. By applying a *Critical Social Science* approach, the researcher was able to conduct a more comprehensive assessment of the possibilities within the context of this study. The distillation process exposed that leadership was at the core of technology adoption in these firms. This was precisely so because of the small autonomous nature of these firms. This introduces another strength of the project, particularly since the literature maintains that owner-managed small firms have largely been ignored in the discourse around technology adoption. It is not uncommon for the literature to represent considerable research using large and small firms. Hence, this present study contributes to the literature in a very important way.

In terms of theory, this study seeks to inform future research on hierarchical technology adoption and distinct leadership typologies. Arguably, this research can inform and influence theories related to technology adoption as well as leadership. The more practical implications relate to the travel and tourism industry particularly to those firms that have a deficient technological infrastructure and are struggling. Given the strength of this work, it is argued that there are fundamental areas that will inform the future of the technology adoption discourse, as they are advanced below:

- From a methodological standpoint, this study challenges the reductionist approaches to technology adoption, as it is outlined in prior studies. Moreover, it reinforces the value of *Critical Social Science* Philosophy in research. Thus, allowing for attendant or subliminal issues to be addressed prior to distillation. Ultimately, it clarifies the position/perspective on critical emergent factors in the research.
- In owner-managed firms, where leadership and ownership overlap, there are unique and unavoidable challenges to innovation. This work acknowledges and accounts for this issue.
- Leadership role is significantly influenced by the size of the firm as well as how simple the management and decision-making structures are.
- Delegating responsibility is rarely achieved by owner-managers since they are personally attached to their firms. This is most evident when key decisions such as the adoption of new technology should be made. Fundamentally, this means that owner-managers will choose between stimulating adoption behavior among staff or restricting adoption behavior based on their unique preferences. Having accounted for these nuances, this work provides a more detailed analysis than previous studies on specific leadership typologies that drive different levels of adoption in firms with similar external characteristics.
- As an original contribution to theoretical foundation, this current study develops leadership typologies at each level of technology adoption. This was never accomplished in previous studies. This development is particularly useful since as the generic transactional and transformational typologies were too deterministic. In that, they

were limited in their application to a hierarchical adoption model. It therefore portrays that ICT adoption and competitiveness is more likely to improve in small owner-managed travel firms, through the adjustment and engagement with one or more of the leadership variables that were identified.

Additionally, this study extends the theory of transformational leadership. While it is acknowledged that generic classifications of leadership have demonstrable value in identifying leaders who are themselves change agents, the findings indicated that the usual classifications lacked explanatory power for how leadership influenced ordered technology adoption along an adoption hierarchy in small, owner-managed firms. Therefore, using the set of classifications advanced in this study, there are deeper considerations for more technologically related variables to better understand the difference in each firms' positionality on the hierarchy. However, this model is only applicable to those firms that are small, owner-managed and operate in a similar context that accounts for the digital divide and culture. Consequently, comparative research across cultures or societies may require different conceptualizations.

The findings of this research is also fundamentally important for practitioners in contexts where the travel agencies that exist have faced commission cuts amidst other serious challenges over the period of 1999–2009. Notwithstanding the changes in the global space and to operational procedures in these firms at the time, the most static area of operations related to the use of technology. This was most evident when there was still very little adoption of new technologies particularly for sales and marketing, even after the adoption of Global Distribution Systems (primarily Sabre and Amadeus) in the 1990s.

Given the noticeable value of leadership, it is argued here that these firms and others that encounter similar challenges should assess the styles of leadership they employ. Since owner-managed firms vary based on technology experience, education, type of industry experience, family composition, risk aversion and intellectual stimulation, it is seemingly more practical for these small firms to improve on variables which can be controlled. For instance, it is suggested that small owner-managed firms can improve the experience with technologies, education levels or the provision of an intellectually stimulating work environment. While this study accounts for the nuances among these firms, it advocates that leaders should be more open to exploring ideas of technological reform for the development of their businesses nonetheless. Ultimately such openness can prepare these firms to become more competitive in the global domain.

There are many areas for future research in this field. It is hoped that this research will encourage more comprehensive research on technology adoption to identify the determinants. It is also anticipated that future research will be more holistic in assessing the determinants of technology adoption. One fundamental area of research that requires deeper exploration is the role of transformational leadership in technology adoption for large firms. Moreover, since context is important, this assessment should occur in a developed country context to identify whether there are key differences, comparatively with developing countries. In this vein, a comparative study across cultures may be insightful in uncovering the influence of culture and digital divide issues. This would be integral to the tourism sector. There is also room for meaningful research to be undertaken on the challenges to leadership in firms that are located in destination countries and firms that are in more developed countries. Finally, to supplement existing qualitative research, future studies should be quantitative in nature to identify other variables and connect to the typologies highlighted in this study in more diverse environments.

References

Aquila-Obra, Ana, and Antonio Padilla-Melendez. 2006. "Organizational Factors Affecting Internet Technology Adoption." *Internet Research* 16(1): 94–110.

Barney, J.B. 2001. "Resource-Based Theories of Competitive Advantage: A Ten Year Retrospective on the Resource-Based View." *Journal of Management* 27: 643.

Bass, Bernard M., Bruce J. Avolio, Dong I. Jung, and Yair Berson. 2003. "Predicting Unit Performance by Assessing Transformational and Transactional Leadership." *Journal of Applied Psychology* 88(2): 207–218.

Daniel, Elizabeth, Hugh Wilson, and Andrew Myers. 2002. "Adoption of E-Commerce by SMEs in the UK: Towards a Stage Model." *International Small Business Journal* 20(3): 253–270.

David, Fred. 2007. *Strategic Management: Concepts and Cases*. Upper Saddle River, NJ: Prentice Hall.

Eastlick, Mary Ann, and Sherry Lotz. 1999. "Profiling Potential Adopters and Non-Adopters of an Interactive Electronic Shopping Medium." *International Journal of Retail and Distribution Management* 27(6): 209–223.

Govers, Robert F., Frank M. Go, and Kumar Kuldeep. 2007. "Promoting Tourism Destination Image." *Journal of Travel Research* 46(1): 15–23.

Kraajenbrink, Jeroen, J. C. Spender, and Aard Groen. 2010. "The Resource-based View: A Review and Assessment of its Critiques." *Journal of Management* 36: 349.

Maurer, Christian, and Veronika Lutz. 2011. "The Impact of Digital Divide on Global Tourism: Strategic Implications of Overcoming Communication Gaps Caused by Digital Inequalities." In *Information and Communication Technologies in Tourism 2011 Enter Proceedings*, edited by Rob Law, Matthias Fuchs, and Francesco Ricci, 265–277. New York: Springer Wein.

Minghetti, Valeria, and Dimitrios Buhalis. 2010. "Digital Divide in Tourism." *Journal of Travel Research* 49(3): 267–281.

Moital, Miguel, Roger Vaughan, and Jonathan Edwards. 2009. "Using Involvement for Segmenting the Adoption of E-commerce in Travel." *Service Industries Journal* 29(5): 723–739.

Peterson, Suzanne J., Kristin Byron Walumbwa, and Jason Myrowitz. 2009. "CEO Positive Psychological Traits, Transformational Leadership, and Firm Performance in High-Technology Start-Up and Established Firms." *Journal of Management* 35: 348–368.

Stonehouse, George, and Brian Snowdon. 2007. "Competitive Advantage Revisited: Michael Porter on Strategy and Competitiveness." *Journal of Management Inquiry* 16(3): 256–273.

Thong, James Y.L., and C. S. Yap. 1995. "CEO Characteristics, Organisational Characteristics and Information Technology Adoption in Small Business." *Omega-international Journal of Management Science* 23 (4): 429–442.

Victorino, Liana, Rohit Verma, Gerhard Plaschka, and Chekitan Dev. 2006. "Service Innovation and Customer Choices in the Hospitality Industry." *Managing Service Quality* 15(6): 555–576.

Appendix

Summary of Online Adoption Models

Moersch (1995)	Six stages: 1. Information seeking 2. Exploration 3. Ecommerce adoption 4. Integration and Content Processing 5. Collaboration 6. Refinement
Burgess and Cooper (1998)	Three stages: 1. Electronic promotion of products 2. Customer interaction 3. Online orders, sales, and payment
Allcock et al. (1999)	Four stages: 1. Threshold (computer introduction) 2. Beginner (limited networks) 3. Intermediate (static websites) 4. Advanced (email, intranet, extranet)
Earl (2000)	Six stages: 1. Homepages (external communication) 2. Internal Communication 3. Online buying and Selling 4. E-business conversion 5. E-enterprise 6. Continuous reinventing

(continued)

A. Spencer, *Technology Adoption in the Caribbean Tourism Industry*, https://doi.org/10.1007/978-3-030-61584-0

(continued)

Heeks (2000)	Four stages: 1. Simple email use 2. Simple website use 3. Online transactions 4. Service Delivery
Mackay et al. (2000)	Six stages: 1. No online presence 2. Static online presence 3. Interactive online presence 4. Complete internet transactions 5. Integration of front and back office 6. Extended enterprise
Willcocks (2000)	Four stages: 1. Web page development 2. Transaction system development 3. Process integration 4. E-business
Wiertz (2001)	Four stages: 1. Access 2. E-procurement 3. Online promotions 4. E-sales
Daniel et al. (2002)	Four stages: 1. Developers (minimal e-commerce) 2. Communicators 3. Web presence 4. Transactors
Levy and Powell (2002)	Four stages: 1. Email and Informative websites 2. Internal and external communication 3. Business Networking 4. Electronic data interchange systems
Rayport and Jaworski (2002)	Three stages: 1. Broadcast (static websites) 2. Internet interaction with customers 3. Online transactions
Rao et al. (2003)	Four stages: 1. Web presence 2. Two-way communication 3. Online transactions 4. Enterprise integration
Chan and Swatman (2004)	Four stages: 1. Ecommerce adoption 2. Centralized e-commerce 3. New technologies 4. Customer satisfaction

(continued)

(continued)

Beck et al. (2005)	Four stages: 1. Online advertising 2. Online sales and after-sales services 3. Online procurement 4. Electronic data interchange systems
Gatautis and Neverauskas (2005)	Three stages: 1. Electronic data interchange systems 2. Centralized e-commerce 3. Global e-commerce
Lefebvrea et al. (2005)	Six stages: 1. Ecommerce non-adopters 2. Non-adopters with interest 3. Electronic content creation 4. Electronic transactions 5. Complex electronic transactions 6. Electronic collaboration
Gandhi (2006)	Four stages: 1. Online promotions 2. Online customer interaction 3. Online ordering and payment 4. Feedback and after-sales services
Al-Qirim (2007)	Two stages: 1. Starters (e-mail and passive websites) 2. Advanced level (website sales)
Chen and McQueen (2008)	Four stages: 1. Information search and email communication 2. Static online marketing websites 3. Online ordering with manual payment 4. Online transactions and invoicing

References

Aaker, D.A. 2009. *Managing Brand Equity*. New York: Simon and Schuster.

Abou-Shouk, M., P. Megicks, and W.M. Lim. 2013. "Perceived Benefits and E-commerce Adoption by SME Travel Agents in Developing Countries: Evidence from Egypt." *Journal of Hospitality & Tourism Research* 37(4): 490–515.

Afuah, A., and C.L. Tucci. 2003. *Internet Business Models and Strategies: Text and Cases* (Vol. 2). New York: McGraw-Hill.

AIM. 2002. *Austrian Internet Monitor 2002*. Vienna: AIM.

———. 2003. *Austrian Internet Monitor 2003*. Vienna: AIM.

Akkermans-Allen, Hans. 2001. "Intelligent E-Business: From Technology to Value." *IEEE Intelligent Systems* 16: 8–11.

Alford, Philip. 2004. "Critical Theory: An Alternative Solution to IT Planning Implementation Problems in Tourism." In *ENTER 11: Information and Communication Technologies in Tourism Conference 2004: Proceedings of the International Conference in Cairo, Egypt*, edited by Andrew J. Frew, 271–282. Vienna, Austria: Springer.

Amit, R., & Zott, C. 2001. "Value Creation in E-business." *Strategic Management Journal* 22(6–7): 493–520.

Anckar, B. 2003. "Consumer Intentions in Terms of Electronic Travel Distribution: Implications for Future Market Structures." *E-Service* 2(2): 68–86.

Anckar, B. and P. Walden. 2001. "Self-Booking of High-and Low-Complexity Travel Products: Exploratory Findings." *Information Technology & Tourism* 4(3–4): 151–165.

Andrews, P.H., and J.E. Baird. 2005. *Communication for Business and the Professions*. Long Grove: Waveland Press Inc.

A. Spencer, *Technology Adoption in the Caribbean Tourism Industry*, https://doi.org/10.1007/978-3-030-61584-0

Aquila-Obra, Ana, and Antonio Padilla-Melendez. 2006. "Organizational Factors Affecting Internet Technology Adoption." *Internet Research* 16(1): 94–110.

Avolio, B.J., W.L. Gardner, F.O. Walumbwa, F. Luthans, and D.R. May. 2004. "Unlocking the Mask: A Look at the Process by Which Authentic Leaders Impact Follower Attitudes and Behaviors." *The Leadership Quarterly* 15(6): 801–823.

Bagozzi, R.P. 2007. "The Legacy of the Technology Acceptance Model and a Proposal for a Paradigm Shift." *Journal of the Association for Information Systems* 8(4): 3.

Baker, Elizabeth, Mary Barrett, and Lesley Roberts. 2002. *Working Communication*. Queensland: John Wiley & Sons.

Baloglu, Seyhmus. 2000. "A Path Analytic Model of Visitation Intention Involving Information Sources, Socio-Psychological Motivations and Destination Image." *Journal of Travel and Tourism Marketing* 8(3): 81–90.

Banff, Canada, Simon Hudson, and Norma Lang. 2002. "A Destination Case Study of Marketing Tourism Online." *Journal of Vacation Marketing* 8: 155.

Barnett, Martin, and Craig Standing. 2001. "Repositioning Travel Agencies on the Internet." *Journal of Vacation Marketing* 7: 143.

Barney, J.B. 2001. "Resource-Based Theories of Competitive Advantage: A Ten Year Retrospective on the Resource-Based View." *Journal of Management* 27: 643.

———. 2014. *Gaining and Sustaining Competitive Advantage*. London: Pearson Higher Education.

Bass, B.M., & Avolio, B.J. 2003. MLQ Multifactor Leadership Questionnaire. Redwood City: Mind Garden. www.mindgarden.com.

Bass, Bernard M., Bruce J. Avolio, Dong I. Jung, and Yair Berson. 2003. "Predicting Unit Performance by Assessing Transformational and Transactional Leadership." *Journal of Applied Psychology* 88(2): 207–218.

Bass, B.M., and R.E. Riggio. 2006. *Transformational Leadership*. London: Psychology Press.

Bear Stearns. 2000. *Point, Click, Trip: An Introduction to the On-line Travel Industry*. Bear Stearns & Co.

Beck, Andrew, Peter Bennett, and Peter Wall. 2004. *Communication Studies: The Essential Resource*. London: Routledge.

Beck, Roman, Rolf Wigand, and Wolfgang Konig. 2005. "The Diffusion and Efficient Use of Electronic Commerce among Small and Medium-Sized Enterprises: An International Three Industry Survey." *Electronic Markets* 15(1): 38–52.

Beckinsale, Martin, Monder Ram, and Nicholas Theodorakopoulos. 2011. "ICT Adoption and Ebusiness Development: Understanding ICT Adoption amongst Ethnic Minority Businesses." *International Small Business Journal* 29: 193–219.

Beekhuyzen, Jenine, Liisa Hellens, and Mark Siedle. 2005. "Cultural Barriers in the Adoption of Emerging Technologies." Accessed March 2, 2010. http://www.ucd.smartinternet.com.au/Documents/Cultural_Barriers.pdf.

Bei, Lien-Ti, Etta Y.I. Chen, and Richard Widdows. 2004. "Consumers' Online Information Search Behavior and the Phenomenon of Search vs. Experience Products." *Journal of Family and Economic Issues* 25(4): 449–467.

Beirne, Elizabeth, and Philip Curry. 2012. "The Impact of the Internet on the Information Search Process and Tourism Decision-Making." In *Information and Communication Technologies in Tourism 1999: Proceedings of the International Conference in Innsbruck*, edited by Dimitrios Buhalis, and Walter Schertler, 88–97. Wien: Springer.

Belch, George E., and Michael A. Belch. 2001. *Advertising and Promotion: An Integrated Marketing Communications Perspective*. New York: McGraw Hill.

Beliveau, Annie, and John B. Garwood. 2001. "The Impact of New Technologies on Information Research and Purchasing Behaviour of Quebec Air Travellers." In *Information and Communication Technologies in Tourism 2001: Proceedings of the International Conference in Montreal, Canada*, edited by Pauline Sheldon, Karl Wober, and Daniel Fesenmaier, 310–316. Wien: Springer.

Bennett, Marion, and Chi-Wen Lai. 2005. "The Impact of the Internet on Travel Agencies in Taiwan." *Journal of Tourism and Hospitality Research* 6(1): 1–17.

Bethapudi, A. 2013. "The Role of ICT in Tourism Industry." *Journal of Applied Economics and Business* 1(4): 67–79.

Bhimani, A., 2006. "Comparative Management Accounting Research: Past Forays and Emerging Frontiers." *Handbooks of Management Accounting Research* 1: 343–363.

Bieger, Thomas, and Christian Laesser. 2001. "Segmenting Travel on the Sourcing of Information." In *Consumer Psychology of Tourism, Hospitality and Leisure*, edited by Josef A. Mazanec, 153–168. Wallingford: CABI.

Bigne, Enrique J., Joaquín Aldas, and Luisa Anderu. 2008. "B2B Services: IT Adoption in Travel Agency Supply Chains." *Journal of Services Marketing* 22(6): 454–464.

Brewer, J., and A. Hunter. 2006. *Foundations of Multimethod Research: Synthesizing Styles*. Thousand Oaks: Sage.

Brouthers, Keith D., Lance E. Brouthers, and Steve Werner. 2008. "Resource-Based Advantages in the International Context." *Journal of Management* 34(2): 189.

Brown, James, Chris Hendry, and Paul Harborne. 2007. "Developing Radical Technology for Sustainable Energy Markets: The Role of New Small Firms." *International Small Business Journal* 25: 603–625.

Buhalis, Dimitrios. 1996. *Information and Telecommunication Technologies as a Strategic Tool for Tourism Enhancement at Destination Regions*. Innsbruck: Springer Verlag.

———. 1997. *Catalysts in Introducing Information Technology in Small and Medium Sized Hospitality Organizations*. Edinburgh: Springer Verlag.

———. 1998. *Strategic Use of Information Technologies in the Tourism Industry*. Elsevier Science.

———. 2000. "Tourism and Information Technologies: Past, Present and Future." *Tourism Recreation Research* 25(1): 41–58.

———. 2001. "Tourism Distribution Channels: Practices, Issues and Processes." In *Tourism Distribution Channels: Practices, Issues and Transformations*, edited by Dimitrios Buhalis and Eric Laws, 7–32. New York: Continuum.

———. 2002. *eTourism: Information Technologies for Strategic Tourism Management*. Essex: Pearson Education.

———. 2004. "E-Airlines: Strategic and Tactical Use of ICT's in the Airline Industry." *Information and Management* 41(7): 805–825.

Buhalis, Dimitrios, and Eric Laws. 2001. *Tourism Distribution Channels: Practices, Issues and Transformations*. London: Continuum.

Buhalis, Dimitrios, and Maria Cristina Licata. 2002. "The Future eTourism Intermediaries." *Journal of Tourism Management* 23(3): 207–220.

Buhalis, Dimitrios, and Ourania Deimezi. 2004. "E-Tourism Developments in Greece: Information Communication Technologies Adoption for Strategic Management of the Greek Tourism Industry." *Tourism and Hospitality Research* 5(2): 1467–3584.

Burgess, L., J.P. Sargent, J. Cooper, and N. Cerpa. 2005. *A Comparative Analysis of the Use of the Web for Destination Marketing by Regional Tourism Organisations in Chile and the Asia Pacific. Collaborative Electronic Commerce Technology and Research*. Chile: Universidad de Talca. Retrived at https://ro.uow.edu.au/infopapers/1599/.

Burns, James M. 1978. *Leadership*. New York: Harper and Row.

Cai, Liping. A., Ruomei Feng, and Deborah Breiter. 2004. "Tourist Purchase Decision Involvement and Information Preferences." *Journal of Vacation Marketing* 10(2): 138–148.

Carland, J.W., F. Hoy, W.R. Boulton, and J.A.C. Carland. 2007. "Differentiating Entrepreneurs from Small Business Owners: A Conceptualization." In *Entrepreneurship*, 73–81. Berlin and Heidelberg: Springer.

Carlsson, S.A. 2005. "Developing Information Systems Design Knowledge: A Critical Realist Perspective." *The Electronic Journal of Business Research Methodology* 3(2): 93–102.

Castells, Manuel. 2000. *End of Millennium*. 2nd ed. Oxford: Blackwell.

Caves, Richard E. 1980. "Industrial Organization, Corporate Strategy and Structure." *Journal of Economic Literature* 18(1): 64–92.

Central Information Technology Office. 2007. "National e-Readiness Status." Accessed May 12, 2011. http://www.cito.gov.jm/content/national-e-readiness-status.

Chan, Caroline, and Paula Swatman. 2004. "B2B E-Commerce Stages of Growth: The Strategic Imperatives." In *37th International Conference on System Sciences*, 1–10. Hawaii.

Chandler, Alfred. 2003. *Strategy and Structure: Chapters in the History of the American Industrial Enterprise*. Fairless Hills: Beard Books.

Chen, I.J. and A. Paulraj. 2004. "Understanding Supply Chain Management: Critical Research and a Theoretical Framework." *International Journal of Production Research* 42(1): 131–163.

Chen, Jihong, and Robert McQueen. 2008. "Factors Affecting E-commerce Stages of Growth in Small Chinese firms in New Zealand: An Analysis of Adoption Motivators and Inhibitors." *Journal of Global Information Management* 16(1): 26–60.

Chen, Joseph S. A. 2000. "Comparison of Information Usage between Business and Leisure Travelers." *Journal of Hospitality and Leisure Marketing* 7(2): 65–75.

Chen, Joseph. S., and Dogan Gursoy. 2000. "Cross-Cultural Comparison of the Information Sources Used by First-time and Repeat Travelers and its Marketing Implications." *International Journal of Hospitality Management* 19: 191–203.

Cheyne, Jo, Mary Downes, and Stephen Legg. 2006. "Travel Agent vs. Internet: What Influences Travel Consumer Choices?" *Journal of Vacation Marketing* 12(1): 41–57.

Choa, Myeong-Hyeon. 1998. "Ownership Structure, Investment and the Corporate Value: An Empirical Analysis." *Journal of Financial Economics* 47(1): 103–121.

Choi, Soon-Yong, and Dale Stahl. 1997. *The Economics of Electronic Commerce*. Indiana: Macmillan Technical Publishing.

Chon, Kye-Sung. 1999. "Tourist Information-Seeking Behaviour and Its Implications." *Hospitality Research Journal* 14: 485–490.

Chorn, Norman. 1994. "The New Logistics Organization." In *The Gower handbook of Logistics and Distribution Management*. 4th ed., edited by John Gattorna, 501–508. Aldershot, England: Gower Publishing.

Christensen, Edward W., and James Bailey. 1997. "A Source Accessibility Effect on Media Selection." *Management Communication Quarterly* 10(3): 373–387.

Christopher, Martin. 1986. *The Strategy of Distribution Management*. Oxford: Heinemann Professional Publishing.

———. 1992. *Logistics and Supply Chain Management: Strategies for Reducing Cost and Improving Service*. London: Prentice Hall.

———. 1998. *Logistics and Supply Chain Management: Strategies for Reducing Cost and Improving Service*. London: Prentice Hall.

Churchill, Gilbert A. 1995. *Marketing Research*. Chicago: Dryden Press.

Clark, Peter, and Richard Blundel. 2007. "Critical Realism and the Evolution of Business Knowledge: A Methodological Reappraisal." *Management and Organizational History* 2(1): 45–62.

Cockburn, Claire, and Thomas Wilson. 1996. "Business Use of the World Wide Web." *International Journal of Information Management* 16(2): 83–102.

Cole, Rebel A., and Hamid Mehran. 1998. "The Effect of Changes in Ownership Structure on Performance: Evidence from the Thrift Industry." *Journal of Financial Economics* 50(3): 291–317.

Colli, A. 2011. "Business History in Family Business Studies: From Neglect to Cooperation? *Journal of Family Business Management* 1: 12.

comScore. 2002. "Online Travel Launches First Quarter 2002 E-commerce Sales to Record $17 Billion Level." comScore Press release. Accessed February 13, 2010. www.comscore.com.

Conlisk, John. 1996. "Why Bounded Rationality?" *Journal of Economic Literature* 34(2): 669–700.

Connolly, Daniel J., Michael D. Olsen, and Richard Moore. 1998. "The Internet as a Distribution Channel." *Cornell Hotel & Restaurant Administration Quarterly* 39(4): 42–54.

Cook, Suzanne. 2001. *Travel Planners: How They Use the Web Today*. Montreal: Springer Verlag.

Cooper, Randolph B., and Robert Zmud. 1990. "Information Technology Implementation Research: A Technological Diffusion Approach." *Management Science* 36(2): 123–139.

Corey, Robert J. 1996. "A Drama-Based Model of Traveler Destination Choice." *Journal of Travel and Tourism Marketing* 5(4): 1–22.

Covin, Jeffery G., and Dennis Slevin. 1988. "The Influence of Organization Structure on the Utility of an Entrepreneurial Top Management Style." *Journal of Management Studies* 25(3): 217–234.

Cox, Andrew. 1998. "Relational Competence and Strategic Procurement Management: Towards an Entrepreneurial and Contractual Theory of the Firm." *European Journal of Purchasing and Supply Management* 2: 57–70.

Cox, Andrew, and Mike Townsend. 1998. *Strategic Procurement in Construction*. London: Thomas Telford.

Crabtree, Benjamin, and William Miller. 1999. "A Template Approach to Text Analysis: Developing and Using Codebooks." In *Doing Qualitative Research*, edited by Benjamin Crabtree, and William Miller, 163–177. Newbury Park, CA: Sage.

Craig, Anita P. 2003. "Culture and the Individual." *Theory and Psychology* 13(5): 629–650.

Cresswell, John W. 2006. *Research Design: Qualitative and Quantitative Approaches*. California: Sage.

Creswell, John W., Vicki L. P. Clark, Michelle L. Gutmann, and William E. Hanson. 2003. "Advanced Mixed Methods Designs." In *Handbook of Mixed Methods in Social and Behavioral Research*, edited by Abbas Tashakkori, and Charles Teddlie, 209–240. London: Sage.

Crotts, John C. 1999. "Consumer Decision Making and Pre-purchase Information Search." In *Consumer Behavior in Travel and Tourism*, edited by Abraham Pizam, and Yoel Mansfeld, 149–168. New York: Haworth Press.

Crotts, John, Abdul Aziz, and Andrew Raschid. 1998. "Antecedents of Supplier's Commitment to Wholesale Buyers in the International Travel Trade." *Tourism Management* 19(2): 127–134.

Cyert, Richard M., and James March. 1963. *A Behavioural Theory of the Firm*. Englewood Cliffs, NJ: Prentice Hall.

Daft, Richard L., and Robert Lengel. 1984. "Information Richness: A New Approach to Managerial Behavior and Organizational Design." *Research in Organizational Behavior* 6: 191–233.

Dahles, Heidi, and Karin Bras. 1999. "Entrepreneurs in Romance: Tourism in Indonesia." *Annals of Tourism Research* 26: 267–293.

Daily, Catherine M., and Marc Dollinger. 1992. "An Empirical Examination of Ownership Structure in Family and Professionally-managed Firms." *Family Business Review* 5(2): 117–136.

Dale, Crispin. 2003. "The Competitive Networks of Tourism E-mediaries: New strategies, New Advantages." *Journal of Vacation Marketing* 9: 109.

Damanpour, Fariborz. 1991. "Organizational Innovation: A Meta-analysis of Effects of Determinants and Moderators." *Academy of Management Journal* 34(3): 555–590.

D'Ambra, John, and Concepcion Wilson. 2004. "Use of the World Wide Web for International Travel: Integrating the Construct of Uncertainty in Information Seeking and the Task- technology Fit (TTF) Model." *Journal of the American Society for Information Science and Technology* 55(8): 731–742.

Dance, Frank E. X. 1967. "A Helical Model of Communication." In *Human Communication Theory*, edited by Frank Dance. New York: Holt, Reinhart and Winston.

Daniel, Elizabeth, Hugh Wilson, and Andrew Myers. 2002. "Adoption of E-Commerce by SMEs in the UK: Towards a Stage Model." *International Small Business Journal* 20(3): 253–270.

Dann, Graham M. 1996. "Tourists' Images of a Destination—An Alternative Analysis." *Journal of Travel and Tourism Marketing* 5(1/2): 41–55.

Danzinger, Claudia, and H. Nowak. 2002. *Internetnutzung und Ecommerce Aktivitäten der österreichischen KMU*. Vienna: Ec3.

David, Fred. 2007. *Strategic Management: Concepts and Cases*. Upper Saddle River, NJ: Prentice Hall.

Davies, Omar. 1986. "An Analysis of the Management of the Jamaican Economy: 1972–1985." *Social and Economic Studies* 35(1): 73–110.

Davis, Fred D. 1989. "Perceived Ease of Use, and User Acceptance of Information Technology." *MIS Quarterly* 13(3): 319–339.

Decrop, Alain. 1999. "Triangulation in Qualitative Tourism Research." *Tourism Management* 20: 157–161.

Deeds, David, Paul Mang, and Michael Fransden. 2004. "The Influence of Firms' and Industries' Legitimacy on the Flow of Capital into High-technology Ventures." *Strategic Organization* 2(1): 9–34.

DeFleur, Melvin L. 1966. *Theories of Mass Communication*. New York: David McKay.

DeHayes, Daniel W., and Robert L. Taylor. 1972. "Making Logistics Work in a Firm." *Business Horizons* 24(1): 1–26.

Dellaert, Benedict G. C. 1999. "The Tourist as a Value Creator on the Internet." In *Information and Communication Technologies in Tourism 1999: Proceedings of the International Conference in Innsbruck, Austria, 1999*, edited by Dimitrios Buhalis, and Walter Schertler, 66–76. Wien: Springer.

DeMartino, Richard, Robert Barbato, and Paul Jacques. 2006. "Exploring the Career/Achievement and Personal Life Orientation Differences between Entrepreneurs and Non Entrepreneurs: The Impact of Sex and Dependents." *Journal of Small Business Management* 44(3): 350–369.

Demsetz, Harold, and Kenneth Lehn. 1985. "The Structure of Corporate Ownership." *Journal of Political Economy* 93(6): 1155–1177.

Denzin, Norman K., and Yvonna Lincoln. 1994. "Entering the Field of Qualitative Research." In *Handbook of Qualitative Research*, edited by Norman Denzin, and Yvonna Lincoln, 1–17. Thousand Oaks.

Desinano, Paolo, and Carlo Vigo. 1994. *Developing Information Technology Options in the Hotel Industry: The Value Chain Approach. Tourism—The State of Art*. New York: John Wiley & Sons.

Dettmer, Harald. 1998. *Tourismuswirtschaft*. Köln: Wirtschaftsverlag Bachem.

Dietrich, Brenda L., Jane L. Snowdon, and JoAnn Washam. 1997. "The Promise of Information Technology in the Travel Industry." In *Information and Communication Technologies in Tourism: Proceedings of the International Conference in Edinburgh, Scotland, 1997*, edited by A. Min Tjoa, 129–139. Wien: Springer.

D'Iribarne, Philippe. 2009. "National Cultures and Organisations in Search of a Theory: An Interpretative Approach." *International Journal of Cross Cultural Management* 9: 309–321.

Dodd, Tim H. 1998. "Influences on Search Behaviour of Industrial Tourists." *Journal of Hospitality and Leisure Marketing* 5(2/3): 77–94.

Dombey, Alyson. 1998. "Separating the Emotion from the Fact: The Effects of the New Intermediaries on Electronic Travel Distribution." In *Information and Communication Technologies in Tourism 1998: Proceedings of the International Conference in Istanbul, Turkey, 1998*, edited by Dimitrios Buhalis, A. Min Tjoa, and Jafar Jafari, 129–138. Wien: Springer.

Downes, Larry, and Chunka Mui. 1998. *Unleashing the Killer App*. Boston: Harvard Business School Press.

Drori, Gili, and Yong Suk Jang. 2003. "The Global Digital Divide: A Sociological Assessment of Trends and Causes." *Social Science Computer Review* 21: 144–161.

Drucker, Peter. 1990. *Post Capitalist Society*. New York: Harper Business.

Duke, Charles R., and Margaret Persia. 1993. "Effects of Distribution Channel Level on Tour Purchasing Attributes and Information Sources." In *Communication and Channel Systems in Tourism Marketing*, edited by Daniel Fesenmaier, and Muzaffer Uysal, 37–55. Gloucestershire: The Hawthorn Press.

Dulle, Frankwell W., and M.K. Minishi-Majanja. 2011. "The Suitability of the Unified Theory of Acceptance and Use of Technology (UTAUT) Model in Open Access Adoption Studies." *Information Development* 27(1): 32–45.

Durand, Rodolphe, and Vicente Vargas. 2003. "Ownership, Organization, and Private Firms' Efficient Use of Resources." *Strategic Management Journal* 24(7): 667–75.

Dyerson, Romano, and G. Harindranath. 2007. "ICT Adoption and Use by SMEs in the UK: A Survey of South East." Paper Presented at the Portland International Center for Management of Engineering and Technology (PICMET), Portland, Oregon, USA.

Eames, Donald, and Gregory Norkus. 1988. "Developing Your Procurement Strategy." *Cornell Hotel and Restaurant Administration Quarterly* 29: 30–33.

Earl, Michael. 2000. "Evolving the E-business." *Business Strategy Review* 11(2): 33–38.

Eason, Ken. 1988. *Information Technology and Organizational Change*. London: Taylor & Francis.

Eastlick, Mary Ann, and Sherry Lotz. 1999. "Profiling Potential Adopters and Non-Adopters of an Interactive Electronic Shopping Medium." *International Journal of Retail and Distribution Management* 27(6): 209–223.

Eastman, Richard. 1996. "The Future of Travel Agents." *Travel Weekly* 29.

Eby, David W., Lisa J. Molnar, and Liping A. Cai. 1999. "Content Preferences for In-vehicle Tourist Information Systems: An Emerging Travel Information Source." *Journal of Hospitality and Leisure Marketing* 6(3): 41–58.

EC. 1999. *ISTAG Report—Orientations for Work Programme 2000 and Beyond*. Brussels: European Commission.

———. 2003. *The European Ebusiness Report 2002/2003*. Brussels: European Commission.

Eisenhardt, K.M., and J.A. Martin, 2000. Dynamic Capabilities: What Are They? *Strategic Management Journal* 21(10–11): 1105–1121.

Elenkov, Detelin, and Ivan Manev. 2005. "Top Management Leadership and Influence on Innovation: The Role of Socio-Cultural Context." *Journal of Management* 31(3): 381–402.

Ellard, Al, and Norma Nickerson. 1999. "Navigating Cyberspace: Information Seeking Strategies and Travel Behavior of Montana Vacation Visitors." TTRA Annual Conference.

Elliott, Dawn R. 2000. "Underdevelopment in Jamaica: An Institutionalist Perspective." *Journal of Economic Issues* 34: 393–401.

Engel, James F., Martin Warshaw, and Thomas Kinnear. 1991. *Promotional Strategy*. 7th ed. Boston: Irwin.

Engel, James F., Roger D. Blackwell, and Paul W. Miniard. 1995. *Consumer Behaviour*. 8th ed. Fort Worth, TX: Dryden.

Enright, Michael J., and James Newton. 2005. "Determinants of Tourism Destination Competitiveness in Asia Pacific: Comprehensiveness and Universality." *Journal of Travel Research* 43: 339–350.

Etzel, Michael J., and Russell G. Wahlers. 1985. "The Use of Requested Promotional Materials by Pleasure Travelers." *Journal of Travel Research* 23: 2–6.

Evans, Philip B., and Thomas S. Wurster. 1997. "Strategy and the New Economics of Information." *Harvard Business Review* 75(5): 70–82.

Falk, Martin. 2008. "Effects of Foreign Ownership on Innovation Activities: Empirical Evidence for Twelve European Countries." *National Institute Economic Review* 204: 85–97.

Fall, Lisa T. 2000. *Segmenting Pleasure Travelers on the Basis of Information Sources Usefulness and Personal Value Importance*. East Lansing: Michigan State University.

Feather, John. 1998. *The Information Society: A Study of Continuity and Change*. 2nd ed. London: Library Association Publishing.

Fesenmaier, Daniel R., and Jiann-Min Jeng. 2000. "Assessing the Structure in the Pleasure Trip Planning Process." *Tourism Analysis* 5(1): 13–29.

Fesenmaier, Daniel R., and Christine Vogt. 1992. "Evaluating the Utility of Touristic Information Sources for Planning Midwest Vacation Travel." *Journal of Travel and Tourism Marketing* 1(2): 1–18.

Fesenmaier, Daniel R., and Stefan Klein. 2000. *Information and Communication Technologies in Tourism*. Barcelona: Springer Verlag.

Fischer, Ronald. 2009. "Where Is Culture in Cross Cultural Research? An Outline of a Multilevel Research Process for Measuring Culture as a Shared Meaning System." *International Journal of Cross Cultural Management* 9: 25–49.

Fishbein, Martin. 1967. *Readings in Attitude Theory and Measurement*. New York: Wiley.

Fishbein, Martin, and Icek Azjen. 1975. *Belief, Attitude, Intention and Behaviour: An Introduction to Theory and Research*. Reading, MA: Addision-Wesley.

———. 1980. "Predicting and Understanding Consumer Behavior: Attitude-behavior Correspondence." In *Understanding Attitudes and Predicting Social Behavior*, edited by Martin Fishbein, and Icek Ajzen, 148–156. Englewood Cliffs: Prentice-Hall.

Flognfeldt, Thor, and Ragnar Nordgreen. 1999. *Information Used for Decision of Making Temporary Stops along Tourist Routes in Norway*. Halifax, Canada: Annual TTRA Conference.

Fodness, Dale, and Brian Murray. 1997. "Tourist Information Search." *Annals of Tourism Research* 24(3): 503–523.

———. 1998. "A Typology of Tourist Information Search Strategies." *Journal of Travel Research* 37(2): 108–119.

———. 1999. "A Model of Tourist Information Search Behaviour." *Journal of Travel Research* 37(3): 220–230.

Fontayne, Cynthia. 1996. *The Connected Marketer: Using the Internet, World Wide Web and On-line Services for Marketing Travel*. Madrid: Tourism and New Information Technologies.

Forman, Andrew M., and Ven Sriram. 1991. "The Depersonalization of Retailing—Its Impact on the "Lonely" Customer." *Journal of Retailing* 67: 226–243.

Forrester. 1998. "European New Media Strategies." Forrester Research 16.

———. 1999. "On-line Retail Strategies." Forrester Research 15.

Freeman, R.E. 2010. *Strategic Management: A Stakeholder Approach*. Cambridge: Cambridge University Press.

Frese, Michael, Marco van Gelderen, and Michael Ombach. 2000. "How to Plan as a Small Scale Business Owner: Psychological Process Characteristics of Action Strategies and Success." *Journal of Small Business Management* 38: 1–18.

Freudenthaler, Johannes, and Wilfried Grossmann. 2002. *Data Warehouse Models for Web-data. EC3*. Vienna: University of Vienna.

Frew, Andrew J. 2000. "Information and Communications Technology Research in the Travel and Tourism Domain: Perspective and Direction." *Journal of Travel Research* 39(2): 136–145.

Frey, Susanne, and Schegg Roland. 2000. *Bench Marketing Internet Use for Marketing of Swiss Hotels*. Innsbruck: Springer Verlag.

Fuchs, Matthias, Wolfram Hopken, Andreas Foger, and Martin Kunz. 2009. "E-Business Readiness, Intensity and Impact: An Austrian Destination Management Organization Study." *Journal of Travel Research* 49(2): 165–178.

Fuchs, Matthias, Wolfram Hopken, Andreas Foger, and Martin Kunz. 2010. "E-Business Readiness, Intensity, and Impact: An Austrian Destination Management Organization Study." *Journal of Travel Research* 49(2): 165–178.

Fulk, Janet, Joseph Schmitz, and Charles W. Steinfeld. 1990. "A Social Influence Model of Technology Use." In *Organizations and Communications Technology*, edited by Janet Fulk, and Charles Steinfeld, 117–140. Newbury Park, CA: Sage Publications.

Gage, Nathaniel. 1989. "The Paradigm War and Their Aftermath: A 'Historical' Sketch Research on Teaching." *Education Researcher* 18: 4–10.

Gallo, Peter J., and Lisa Jones Christensen. 2011. "Firm Size Matters: An Empirical Investigation of Organizational Size and Ownership on Sustainability-Related Behaviors." *Business Society* 50: 315–349.

Gandhi, Sunil. 2006. "E-Commerce and Information Technology Act, 2000." *Vidyasagar University Journal of Commerce* 11: 82–91.

Gartner Group. 2000. "Gartner Group's Online Travel Forecast." Accessed October 4, 2010. http://gartner12.gartnerWeb.com/public/static/aboutgg/pressrel/pr010600.ht.

Gatautis, Rimantas, and Bronius Neverauskas. 2005. "E-commerce Adoption in Transition Economies: SMEs Perspectives in Lithuania." In *7th International Conference on Electronic Commerce*, 109–113. Xi'an, China: ACM New York, NY, USA.

Gattorna, John. 1994. *The Gower Handbook of Logistics and Distribution Management*. 4th ed. Aldershot, England: Gower Publishing.

Gatty, Bob. 1998. "E-Commerce Gives Lodging Industry Edge." *Hotel and Motel Management* 213.

Gelade, Garry, Paul Dobson, and Patrick Gilbert. 2006. "National Differences in Organizational Commitment: Effect of Economy, Product of Personality, or Consequence of Culture?" *Journal of Cross-Cultural Psychology* 37(5): 542–556.

Gentner, D., B. Stelzer, and L. Brecht. 2017, June. "Analyzing Customers' Readiness for Digital B2B Business Models." In *ISPIM Innovation Symposium*, 1. The International Society for Professional Innovation Management (ISPIM).

George, Gerard, Johan Wiklund, and Shaker Zahra. 2005. "Ownership and the Internationalization of Small Firms." *Journal of Management* 31(2): 210–233.

Ghobadian, Abby, and Nicholas O'Regan. 2006. "The Impact of Ownership on Small Firm Behaviour and Performance." *International Small Business Journal* 24(6): 555–584.

Ghosh, Shikhar. 1998. "Making Business Sense of the Internet." *Harvard Business Review* 124–135.

Gibson, Cristina, and Dana McDaniel. 2010. "Moving Beyond Conventional Wisdom Advancements in Cross-Cultural Theories of Leadership, Conflict, and Teams." *Perspectives on Psychological Science* 5: 450–462.

Gitleson, Richard J., and John Crompton. 1983. "The Planning Horizons and Sources of Information Used by Leisure Travelers." *Journal of Travel Research* 32: 2–6.

Glaser, Barney G., and Anselm Strauss. 1967. *The Discovery of Grounded Theory: Strategy for Qualitative Research*. New York: Aldine De Gruyter.

Gluck, Frederick W. 1982. "Strategic Management for Competitive Advantage." *Journal of Business Strategy* 2(3): 9–21.

Goeldner, Charles R., and J.R. Brent Ritchie. 2003. *Tourism: Principles, Practices, Philosophies*. 9th ed. Hoboken, NJ: Wiley.

Golafshani, Nahid. 2003. "Understanding Reliability and Validity in Qualitative Research." *The Qualitative Report* 8(4): 597–607.

Golding, Peter, and Graham Murdock. 2001. "Digital Divide: Communications Policy and its Contradictions." *New Economy* 8(2): 110–115.

Goldsmith, Ronald E., and Stephen Litvin. 1999. "Heavy Users of Travel Agents: A Segmentation Analysis of Vacation Travelers." *Journal of Travel Research* 38: 127.

Goossens, Cees F. 1994. "External Information Search: Effects of Tour Brochures with Experiential Information." In *Economy Psychology of Travel and Tourism*, edited by John Crotts, and W. Fred Van Raaij, 89–107. New York: The Haworth Press.

———. 2000. "Tourism Information and Pleasure Motivation." *Annals of Tourism Research* 27(2): 301–321.

Gordijn, Jaap, and Hans Akkermans. 2001. "Designing and Evaluating E-business Models." *IEEE Intelligent Systems* 16(4): 11–17.

Govers, Robert F. 2004. "Projected Destination Image on the World Wide Web: Content Analysis of Online Textual Representations of Dubai." Conference Proceedings for Information and Communication Technologies in Tourism.

Govers, Robert F., Frank M. Go, and Kumar Kuldeep. 2007. "Promoting Tourism Destination Image." *Journal of Travel Research* 46(1): 15–23.

Graeupl, Alice, and Scott McCabe. 2003. "Use of the Internet by 'Silver Surfers' for Travel and Tourism Information Use and Decision-Making: A Conceptual Review and Discussion of Findings." In *Information and Communication Technologies in Tourism*, edited by Andrew Frew, Martin Hitz, and Peter O'Connor, 144–155. New York: Springer.

Grandon, Elizabeth, and J. Michael Pearson. 2004. "Electronic Commerce Adoption: An Empirical Study of Small and Medium US Businesses." *Information and Management* 42: 197–216.

Grant, Robert M. 1995. *Contemporary Strategy Analysis*. Oxford: Blackwell.

Gray, Colin. 2006. "Stage Models of ICT Adoption in Small Firms." In *Impact of E-commerce on Consumers and Small Firms*, edited by Salvatore Zappala, and Colin Gray, 3–20. Aldershot: Ashgate.

Greene, Jennifer C., and Valerie J. Caracelli. 2003. "Making Paradigmatic Sense of Mixed Methods Practice." In *Handbook of Mixed Methods in Social and Behavioral Research*, edited by Charles Teddlie, and Abbas Tashakkori, 91–110. London: Sage.

Greenwood, Royston, David Deephouse, and Stan Li. 2007. "Ownership and Performance of Professional Service Firms." *Organization Studies* 28(2): 219–238.

Greenwood, Royston, and Laura Empson. 2003. "The Professional Partnership: Relic or Exemplary Form of Governance?" *Organization Studies* 24(6): 909–933.

Grennes, Thomas. 1997. *Introduction to Scientific Theory and Method.* Oslo: Tano Aschehoug.

Grieger, M. 2003. "Electronic Marketplaces: A Literature Review and a Call for Supply Chain Management Research." *European Journal of Operational Research* 144(2): 280–294.

Grimm, Laurence G., and Paul R. Yarnold. 1995. *Reading and Understanding Multivariate Statistics*. Washington, DC: American Psychological Association.

Grönroos, Christian. 1993a. "Toward a Third Phase in Service Quality Research." In *Advances in Services Marketing and Management*, edited by Teresa A. Swartz, David A. Bowen, and Stephen W. Brown, 2: 49–64. Greenwich, CT: JAI Press.

Grönroos, C. 1993b. "An applied Service Marketing Theory." *European Journal of Marketing* 16(7): 30–34.

Grossmann, Wilfried, and Marcus Hudec. 2003. "Web Usage Mining in Ecommerce." *International Journal of eBusiness* (Special Issue EC3).

Guba, Egon G. 1987. "Naturalistic Evaluation." *New Directions for Program Evaluation* 34: 23–43.

Guillén, Mauro, and Sandra Suárez. 2005. "Explaining the Global Digital Divide: Economic, Political and Sociological Drivers of Cross-National Internet Use." *Social Forces* 84(2): 681–708.

Gunkel, David. 2003. "Second Thoughts: Toward a Critique of the Digital Divide." *New Media and Society* 5(4): 499–522.

Gursoy, Dogan. 2003. "Prior Product Knowledge and Its Influence on the Traveler's Information Search Behaviour." *Journal of Hospitality and Leisure Marketing* 10(3/4): 113–130.

Gursoy, Dogan, and Joseph Chen. 2000. "Competitive Analysis of Cross Cultural Information Search Behavior." *Tourism Management* 21: 583-590.

Gursoy, Dogan, and Ken W. McCleary. 2004. "An Integrative Model of Tourists' Information Search Behavior." *Annals of Tourism Research* 31(2): 353–373.

Gyamfi, Alexander. 2005. "Closing the Digital Divide in Sub-Saharan Africa: Meeting the Challenges of the Information Age." *Information Development* 21(1): 22–30.

Haertsch, Patrick. 2000. *Wettbewerbstrategien für Electronic Commerce: Eine Kritische Überprüfung Klassischer Strategiekonzepte*. Lohmar; Köln, Josef Eul Verlag.

Hair, Joseph F., Rolph E. Anderson, Ronald L. Tatham, and William Black. 1998. *Multivariate Data Analysis*. 5th ed. Upper Saddle River, NJ: Prentice-Hall.

Hamel, Gary, and C. K. Prahalad. 1994. *Competing for the Future*. Boston: Harvard Business School Press.

Hamlin, Robert, and Jenny Sawyer. 2007. "Developing Effective Leadership Behaviours: The Value of Evidence-based Management." *Business Leadership Review* 4(4): 1–16.

Hanssler, Karl Heinz. 2001. *Management in der Hotellerie und Gastronomie*. München: Oldenburg Verla.

Haralambos, Michael, and Martin Holborn. 1980. *Sociology Themes and Perspectives*. Collins Educational.

Haralambos, M., and M. Holborn. 2008. *Sociology: Themes and Perspectives*. UK: HarperCollins.

Harland, Christine M. 1996. "Supply Chain Management, Purchasing and Supply Management, Logistics, Vertical Integration, Materials Management and Supply Chain Dynamics." In *Blackwell Encyclopedic Dictionary of Operations Management*, edited by Nigel Slack. Oxford, UK: Blackwell.

Harrison, Graeme, and Jill McKinnon. 1999. "Cross-Cultural Research in Management Control Systems Design: A Review of the Current State." *Accounting, Organizations and Society* 24: 483–506.

Hatton, Gerry. 1994. "Designing a Warehouse or Distribution Centre." In *The Gower Handbook of Logistics and Distribution Management*, edited by John Gattorna, 175–193. 4th ed. Aldershot, England: Gower Publishing.

Hatton, Mike. 2004. "Redefining the Relationships: The Future of Travel Agencies and the Global Agency Contract in a Changing Distribution System." *Journal of Vacation Marketing* 10: 101.

Hedna. 2002. "Despite Challenges in 2001: HEDNA Reports that GDS Bookings for Hotels Remain Strong." London: Hedna Press Release. Accessed January 2, 2010. www.hedna.org.

Heeks, R. 2000. "Analysing E-commerce for Development." Accessed April 7, 2009. http://www.sed.manchester.ac.uk/idpm/research/publications/wp/di/short/di_sp04.pdf.

Heinich, Nathalie. 2010. "What Does 'Sociology of Culture' Mean? Notes on a Few Trans-Cultural Misunderstandings." *Cultural Sociology* 4: 257–265.

Heung, Vincent C. S. 2003. "Barriers to Implementing E-Commerce in the Travel Industry: A Practical Perspective." *International Journal of Hospitality Management* 22(1): 111-118.

Hitt, Michael A., R. Duane Ireland, and Robert E. Hoskisson. 2001. *Strategic Management: Competitiveness and Globalization*. 4th ed. Mason, OH: South Western Publishing.

HoCo. 2000. *Worldwide Hotel Industry Study, 1996–2000*. London: Horwath Consulting.

Hoffman, Donna, and Thomas P. Novak. 1996. "Marketing in Hypermedia Computer-mediated Environments: Conceptual Foundations." *Journal of Marketing* 60: 50–68.

Hoffman, Eric, and Paul Roman. 1984. "Information Diffusion in the Implementation of Innovation Process." *Communication Research* 11(1): 117–140.

Hofstede, Geert. 1980. *Culture's Consequences: International Differences in Work-related Values*. Beverly Hills, CA: Sage.

———. 1999. *Cultures and Organizations: Software of the Mind*. Maidenhead: McGraw-Hill.

Hofstede, Geert, and Michael H. Bond. 1984. "Hofstede's Culture Dimensions: An Independent Validation Using Rokeach's Value Survey." *Journal of Cross-Cultural Psychology* 15: 417–433.

Hollander, Edwin P. 1993. "Legitimacy, Power and Influence: A Perspective on Relational Features of Leadership." In *Leadership Theory and Research: Perspectives and Directions*, edited by Martin M. Chemers, and Roya Ayman, 29–48. San Diego: Academic Press.

Holloway, J. Christopher. 1994. *The Business of Tourism*. London: Pitman Publishing.

Van den Hooff, Bart, Jasper Groot, and Sander de Jonge. 2005. "Situational Influences on the Use of Communication Technologies: A Meta-Analysis and Exploratory Study." *Journal of Business Communication (1973)* 4(2): 4–27.

Houghton, Kathryn, and Heidi Winklehofer. 2004. "The Effect of Website and E-commerce Adoption on the Relationship between SMEs and Their Export Intermediaries." *International Small Business Journal* 22: 369.

Howard, John A., and Jagdish N. Seth. 1969. *The Theory of Buyer Behaviour*. New York: John Wiley.

Howe, Kenneth. 1988. "Against the Quantitative-Qualitative Incompatibility Thesis." *Educational Researcher* 17(8): 10–16.

Hsieh, Sheauhsing, and Joseph T. O'Leary. 1993. "Communication Channels to Segment Pleasure Travelers." In *Communication and Channel Systems in Tourism Marketing*, edited by Muzaffa Uysal, and Daniel Fesenmaier, 57–75. Gloucestershire: The Hawthorn Press.

Huang, L., and H.T. Tsai. 2003. "The Study of Senior Traveler Behavior in Taiwan." *Tourism Management* 24(5): 561–574.

Huber, George, and Reuben McDaniel. 1986. "The Decision-Making Paradigm of Organizational Design." *Management Science* 32(5): 572–589.

Hyde, Kenneth F. 2000. "A Hedonic Perspective on Independent Vacation Planning, Decision-Making and Behaviour." In *Consumer Psychology of Tourism, Hospitality and Leisure*, edited by Arch G. Woodside, 177–191. New York: CABI.

Inkpen, Gary. 1998. *Information Technology for Travel and Tourism*. Edinburgh: Addison Wesley.

Jamaica Gleaner. 2000. "Travel Agents to Boycott AA." Accessed November 14, 2010. http://jamaica-gleaner.com/gleaner/20000111/f1.html.

———. 2011. "Tourism Minister Targets Illegal Travel Agents." Accessed July 27, 2011. http://jamaica-gleaner.com/gleaner/20110724/business/business7.html.

James, Jeffrey. 2004. "The Global Digital Divide in the Internet: Developed Countries Constructs and Third World Realities." *Journal of Information Science* 31(2): 114–123.

Jang, SooCheong. 2004. "The Past, Present, and Future Research of Online Information Search." *Journal of Travel and Tourism Marketing* 17(2/3): 41–47.

Jaskyte, Kristina, and Kisieliene Audrone. 2006. "Organizational Innovation: Comparison of Nonprofit Human-Service Organizations in Lithuania and the United States." *International Social Work* 49(2): 165–176.

Jeng, Jiann-Min, and Daniel R. Fesenmaier. 1999. "Understanding Tourist Information Search Patterns: Insight for Developing Decision Appliances." TTRA Annual Conference.

Jenkins, Roger L. 1978. "Family Vacation Decision-Making." *Journal of Travel Research* 16(4): 2–7.

Jennings, Gayle. 2001. *Tourism Research*. Milton, Australia: John Wiley & Sons.

Jung, Timothy H., and Richard Butler. 2000. "The Measurement of the Marketing Effectiveness of the Internet in the Tourism and Hospitality Industry." In *Information and Communication Technologies in Tourism 2000: Proceedings of the International Conference in Barcelona, Spain*, edited by Daniel Fesenmaier, Stefan Klein, and Dimitrios Buhalis, 460–472. Wien: Springer.

Jung, Timothy H., Panos Louviers, and Harmen Oppewal. 2002. "Channel Management Strategy in the E-Commerce Environment: A Portfolio Management Approach." In *Information and Communications Technologies in Tourism*, edited by Karl Wober, Andrew Frew, and Martin Hitz, 17–26. New York: Springer Verlag.

Kadison, Maria. 1998. "On-line Retail Strategies." Forrester Research 15.

Kalakoa, Ravi, and Andrew B. Whinston. 1996. *Frontiers of eCommerce*. Reading: Addison-Wesley.

Karamustafa, Kurtulus. 2000. "Marketing-Channel Relationships: Turkey's Resort Purveyors' Interactions with International Tour Operators." *Cornell Hotel and Restaurant Administration Quarterly* 41(4): 21–32.

Karcher, Karsten. 1997. *Reinventing the Package Holiday Business*. Wiesbaden: Deutscher Universitaetsverlag.

Karnes, Frances, and Jane Chauvin. 1985. *Leadership Skills Inventory Forms*. Scottsdale: Great Potential Press.

Keller, Kenneth. 2008. "From Here to There in Information Technology: The Complexities of Innovation." *American Behavioral Scientist* 51(1): 97–106.

Kelman, Herbert C. 1961. "Process of Opinion Change." *Public Opinion Quarterly* 25: 57–78.

Kemp, Stephen. 2005. "Critical Realism and the Limits of Philosophy." *European Journal of Social Theory* 8: 171–191.

Kerr, Barbara A. 1988. *Review of the Leadership Skills Inventory*. Iowa City: The University of Iowa.

Kerstetter, Deborah, and Mi-Hea Cho. 2004. "Prior Knowledge, Credibility and Information Search." *Annals of Tourism Research* 31(4): 961–985.

Kim, Eonsoo, Nam Dae-Il, and J.L. Stimpert. 2004. "The Applicability of Porter's Generic Strategies in the Digital Age: Assumptions, Conjectures, and Suggestions." *Journal of Management* 30(5): 569–589.

King, William R. 1998. "IT Enhanced Productivity and Profitability." Information System Management 15: 64-68.

Kingsley, Ilana, and Daniel R. Fesenmaier. 1995. "Travel Information Kiosks: An Emerging Communications Channel for the Tourism Industry." *Journal of Travel and Tourism Marketing* 4(1): 57–70.

Kirkman, Geoffrey, Peter Cornelius, Jeffrey Sachs, and Klaus Schwab. 2002. *Global Information Technology Report 2001–2002*. Oxford: Oxford University Press.

Klaus, Peter G. 1985. "Quality Epiphenomenon: The Conceptual Understanding of Quality in Face-to-face Service Encounters." In *The Service Encounter: Managing Employee / Customer Interaction in Service Businesses*, edited by John A. Czepiel, Michael R. Solomon, and Carol F. Surprenant, 17–33. Lexington, MA: Lexington Books.

Klein, Barbara D. 2001. "User Perceptions of Data Quality: Internet and Traditional Text Sources." *The Journal of Computer Information Systems* 41(4): 9–15.

Klein, Stefan. 1996. *Interorganisationssysteme und Unternehmensnetzwerke*. St. Gallen: Universität St. Gallen.

———. 2002. *Web Impact of the Distribution Structure for Flight Tickets*. Innsbruck: Springer Verlag.

Klein, Stefan, and Beat Schmid. 1996. *Information and Communication Technologies in Tourism*. Innsbruck, Springer Verlag.

Klein, Stefan, and S. Güler. 1999. *Hotel Reservation Systems on the Internet: Custom Design vs. Standard Software*. Innsbruck: Springer Verlag.

Kotler, Philip, and Gary Armstrong. 2008. *Principles of Marketing*. 12th ed. Upper Saddle River, NJ: Pearson-Prentice Hall.

Koufaris, M., 2002. "Applying the Technology Acceptance Model and Flow Theory to Online Consumer Behavior." *Information Systems Research* 13(2): 205–223.

Kouzes, James M., and Barry Posner. 1987. *The Leadership Challenge: How to Keep Getting Extraordinary Things Done in Organizations*. San Francisco, CA: Jossey Bass.

Kowalewski, Oskar, Oleksandr Talavera, and Ivan Stetsyuk. 2010. "Influence of Family Involvement in Management and Ownership on Firm Performance: Evidence from Poland." *Family Business Review* 23(1): 45–59.

Kraajenbrink, Jeroen, J. C. Spender, and Aard Groen. 2010. "The Resource-Based View: A Review and Assessment of Its Critiques." *Journal of Management* 36: 349.

Krippendorff, Klaus. 1989. Shannon, Claude. In *International Encyclopedia of Communication*, edited by Erik Barnouw, 59–61. New York: Oxford University Press.

Kuan, Kevin K. Y., and Patrick Chau. 2001. "A Perception-based Model for EDI Adoption in Small Businesses Using Technology-Organization-Environment Framework." *Information and Management* 38: 507–521.

Kuhnert, Karl, and Philip Lewis. 1987. "Transactional and Transformational Leadership: A Constructive/Developmental Analysis." *The Academy of Management Review* 12(4): 648–657.

Kulkarni, D. 2019. "Virtual Shopping Cart Abandonment: A Study of Motives and Demographics." Available at SSRN 3405973.

Kwon, Winston, Ian Clarke, and Ruth Wodak. 2009. "Organizational Decision-Making, Discourse, and Power: Integrating across Contexts and Scales." *Discourse and Communication* 3: 273–302.

Lang, Tania. 2000. "The Effect of the Internet on Travel Consumer Purchasing Behaviour and Implications for Travel Agencies." *Journal of Vacation Marketing* 6: 368.

Langley, Ann, Henry Mintzberg, Patricia Pitcher, Elizabeth Posada, and Jan Saint-Macary. 1995. "Opening Up Decision Making: The View from the Black Stool." *Organisation Science* 6(3): 260–279.

Larsen, Tor, and Eugene McGuire. 1998. *Information Systems Innovation and Diffusion: Issues and Directions*. Hershey, PA: Idea Group Publishing.

Lash, Scott. 1994. "Reflexivity and Its Doubles: Structure, Aesthetics, Community." In *Reflexive Modernization*, edited by Ulrich Beck, Anthony Giddens, and Scott Lash, 110–173. Cambridge: Polity Press.

Lasswell, Harold D. 1948. "The Structure and Function of Communication in Society." In *The Communication of Ideas: A Series of Addresses*, edited by Lyman Bryson, 215–228. New York: Harper and Brothers.

Latzer, Michael. 2009. "Information and Communication Technology Innovations: Radical and Disruptive?" *New Media Society* 11(4): 599–619.

Law, Rob, Kenith Leung, and James Wong. 2004. "The Impact of the Internet on Travel Agencies." *International Journal of Contemporary Hospitality Management* 16(2): 100–107.

Law, Rob, and Rita Leung. 2000. "A Study of Airlines' Online Reservation Services on the Internet." *Journal of Travel Research* 39: 202.

Laws, Eric. 2001. "Distribution Channel Analysis for Leisure Travel." In *Tourism Distribution Channels: Practices, Issues and Transformations*, edited by Dimitrios Buhalis, and Eric Laws, 53–72. London: Continuum.

Laws, Eric, and Dimitrios Buhalis. 2001. "Agendas for Future Research." In *Tourism Distribution Channels: Practices, Issues and Transformations*, edited by Eric Laws, and Dimitrios Buhalis, 371–376. London: Continuum.

Lederer, A. L., Maupin, D. J., Sena, M. P., & Zhuang, Y. 1998, June. "The Role of Ease of Use, Usefulness and Attitude in the Prediction of World Wide Web Usage." In *Proceedings of the 1998 ACM SIGCPR Conference on Computer Personnel Research*, 195–204.

Lee, Steven W. 1988. *Review of the Leadership Skills Inventory*. Terre Haute, IN: Indiana State University.

Lefebvrea, Louis, Elisabeth Lefebvrea, Elie Elia, and Harold Boeck. 2005. "Exploring B-to-B e-Commerce Adoption Trajectories in Manufacturing SMEs." *Technovation* 25: 1443–1456.

Legris, Paul, John Ingham, and Pierre Collerette. 2003. "Why Do People Use Information Technology? A Critical Review of the Technology Acceptance Model." *Information and Management* 40: 191–204.

Levy, Margi, and Philip Powell. 2002. "SME Internet Adoption: Towards a Transporter Model." In *15th Bled Electronic Commerce Conference eReality: Constructing the eEconomy*, 507–521. Bled, Slovenia.

Lewis, Gordon K. 2004. *The Growth of the Modern West Indies*. Kingston: Ian Randle publishers.

Lewis, Barbara, and Pamela McCann. 2004. "Service Failure and Recovery: Evidence from the Hotel Industry." *International Journal of Contemporary Hospitality Management* 16(1): 6–17.

Li, X.R., C.K. Cheng, H. Kim, and J.F. Petrick. 2008. "A Systematic Comparison of First-Time and Repeat Visitors Via a Two-Phase Online Survey." *Tourism Management* 29(2): 278–293.

Liikanen, Erkki. 2002. *European Tourism Forum 2002*. Brussels: European Commission.

Lin, Pei-Jung, Eleri Jones, and Sheena Westwood. 2009. "Perceived Risk and Risk-relievers in Online Travel Purchase Intentions." *Journal of Hospitality Marketing and Management* 18(8): 782–810.

Lincoln, Yvonna S., and Egon G. Guba. 1985. *Naturalistic Inquiry*. Thousand Oaks, CA: Sage.

Lippert, Susan K., and Miles Davis. 2006. "A Conceptual Model Integrating Trust into Planned Change Activities to Enhance Technology Adoption Behaviour." *Journal of Information Science* 32(5): 434–448.

Littlejohn, Stephen W. 1989. *Theories of Human Communication*. 3rd ed. Belmont, CA: Wadsworth.

Lo, Ada, Catherine Cheung, and Rob Law. 2002. "Information Search Behavior of Hong Kong's Inbound Travelers: A Comparison of Business and Leisure Travelers." *Journal of Travel and Tourism Marketing* 13(3): 61–81.

———. 2004. "Information Search Behavior of Mainland Chinese Air Travelers to Hong Kong." *Journal of Travel and Tourism Marketing* 16(1): 43–51.

Looney, Robert E. 1987. *The Jamaican Economy in the 1980s: Economic Decline and Structural Adjustment*. Boulder: Westview Press.

Loudon, David L., and Albert J. Della Bitta. 1993. *Consumer Behaviour: Concepts and Applications*. 4th ed. New York: McGraw-Hill.

Lumsdon, Les M., and Jonathan S. Swift. 1999. "The Role of Tour Operators in South America: Argentina, Chile, Paraguay and Uraguay." *International Journal of Tourism Research* 4(1): 429–439.

Luo, Man, Ruomei Feng, and Liping A. Cai. 2004. "Information Search Behaviour and Tourist Characteristics: The Internet Vis-a-Vis other Information Sources." *Journal of Travel and Tourism Marketing* 17(2/3): 15–25.

Lynskey, Michael. 2004. "Determinants of Innovative Activity in Japanese Technology-Based Start-up Firms." *International Small Business Journal* 22(2): 159–196.

Mahoney, Joseph T., and J. Rajendran Pandian. 1992. "The Resource-Based View Within the Conversation of Strategic Management." *Strategic Management Journal* 13: 363–380.

Mai, J.E. 2016. *Looking for Information: A Survey of Research on Information Seeking, Needs, and Behavior*. Bingley: Emerald Group Publishing.

Mandell, Barbara, and Shilpa Pherwani. 2003. "Relationship between Emotional Intelligence and Transformational Leadership Style: A Gender Comparison." *Journal of Business and Psychology* 17(3): 387–404.

Manfredo, Michael J. 1989. "An Investigation of the Basis for External Information Search in Recreation and Tourism." *Leisure Sciences* 11: 29–45.

Mansell, Robin. 2002. "From Digital Divides to Digital Entitlements in Knowledge Societies." *Current Sociology* 50(3): 407–426.

Mansfeld, Yoel. 1992. "From Motivation to Travel." *Annals of Tourism Research* 19: 399–419.

Marcussen, Carl H. 2002. "Trends in European Internet Distribution: Of Travel and Tourism Services." Accessed April 12, 2010. http://www.crt.dk/uk/staff/chm/trends.htm.

Markwell, Kevin W. 1997. "Dimensions of Photography in a Nature-Based Tour." *Annals of Tourism Research* 24: 131–155.

Marler, Janet H., Xiaoya Liang., and James Hamilton Dulebhon. 2006. "Training and Effective Employee Information Technology Use." *Journal of Management* 32(5): 721–743.

Lord Marshall. 2000. "On Line or On the Heap?" *Journal of Vacation Marketing* 6(4): 292–298.

Marshall, Catherine, and Gretchen B. Rossman. 1995. *Designing Qualitative Research*. Thousand Oaks, CA: Sage.

Maser, Birgit, and Klaus Weiermair. 1998. "Travel Decision-Making: From the Vantage Point of Perceived Risk and Information Preferences." *Journal of Travel and Tourism Marketing* 7(4): 107–121.

Masschelein, Lucas, and Alain Van Buyten. 2002. *Structure, Performance and Competitiveness of European Tourism and Its Enterprises*. London: Price Waterhouse Coopers.

Mathieson, Alister, and Geoffrey Wall. 1982. *Tourism: Economic, Physical and Social Impacts*. London: Longman.

Maurer, Christian, and Veronika Lutz. 2011. "The Impact of Digital Divide on Global Tourism: Strategic Implications of Overcoming Communication Gaps caused by Digital Inequalities." In *Information and Communication Technologies in Tourism 2011 Enter Proceedings*, edited by Rob Law, Matthias Fuchs, and Francesco Ricci, 265–277. New York: Springer Wein.

Maxwell, Joseph A., and Diane M. Loomis. 2003. "Mixed Methods Design: An Alternative Approach." In *Handbook of Mixed Methods in Social and Behavioral Research*, edited by Charles Teddlie, and Abbas Tashakkori, 241–272. London: Sage.

May, Christopher. 2002. *Information Society: A Sceptical View*. Cambridge: Polity Press.

May, Paul. 2001. *Mobile Commerce: Opportunities, Applications, and Technologies of Mobile Business*. Cambridge: Cambridge University Press.

Maykut, Pamela. 1994. *Beginning Qualitative Research: A Philosophical and Practical Guide*. Washington, DC: The Falmer Press.

McIntosh, Alison. 1998. "Mixing Methods: Putting the Tourist at the Forefront of Tourism Research." *Tourism Analysis* 3: 121–127.

Mckay, Judith, Peter Marshall, and Adi Prananto. 2000. "Stages of Maturity for e-Business: The SOG-emodel." Accessed December 3, 2009. http://www.pacis-net.org/file/2000/29-43.pdf.

McLean, Laird D. 2005. "Organizational Culture's Influence on Creativity and Innovation: A Review of the Literature and Implications for Human Resource Development." *Advances in Developing Human Resources* 7(2): 226–246.

McQuail, Denis, and Swen Windahl. 1981. *Communication Models—For the Study of Mass Communications*. London: Longman.

McSweeney, Brendan. 2002. "Hofstede's Model of National Cultural Differences and the Consequences: A Triumph of Faith—A Failure of Analysis." *Human Relations* 55: 89–118.

Meditz, Sandra W., and Dennis Hanratty. 1987. "Caribbean Islands: A Country Study." Washington: GPO for the Library of Congress. Accessed March 14, 2011. http://countrystudies.us/caribbean-islands/24.htm.

Menard, Scott. 1995. *Applied Logistic Regression Analysis*. Thousand Oaks, CA: Sage.

Mercer, Jonathan. 2005. "Rationality and Psychology in International Politics." *International Organization* 59(1): 77–106.

Mill, Robert C., and Alastair Morrison. 1992. *The Tourism System: An Introductory Text*. 2nd ed. Englewood Cliffs, NJ: Prentice Hall.

Minghetti, Valeria, and Dimitrios Buhalis. 2010. "Digital Divide in Tourism." *Journal of Travel Research* 49(3): 267–281.

Mitchell, Susan. 1998. *American Generations: Who They Are, How They Live, What They Think*. 2nd ed. Ithaca, NY: New Strategist.

Mochrie, Robert, Laura Galloway, and David Deakins. 2006. "The Value of Internet Forums to Small Rural Businesses." In *Impact of e-Commerce on Consumers and Small Firms*, edited by Salvatore Zappala, and Colin Gray, 81–100. Aldershot: Ashgate.

Moersch, Christopher. 1995. "Levels of Technology Implementation (loti): A Framework for Measuring Classroom Technology Use." Accessed February 5, 2010. http://loticonnection.com/pfd/LoTiFrameworkNov95.pdf.

Moital, Miguel, Roger Vaughan, and Jonathan Edwards. 2009. "Using Involvement for Segmenting the Adoption of E-commerce in Travel." *Service Industries Journal* 29(5): 723–739.

Money, R. Bruce, and John C. Crotts. 2003. "The Effect of Uncertainty Avoidance on Information Search, Planning, and Purchases of International Travel Vacations." *Tourism Management* 24: 191–202.

Montgomery, Cynthia A., and Sam Harihan. 1991. "Diversified Expansion by Large Established Firms." *Journal of Economic Behaviour* 52(1): 71–89.

Moorthy, Sridhar, Brian T. Ratchford, and Debabrata Talukdar. 1997. "Consumer Information Search Revisited: Theory and Empirical Analysis." *Journal of Consumer Research* 23: 263–277.

Morgan, Nigel, Annette Pritchard, and Sara Abbott. 2001. "Consumer, Travel and Technology: A Bright Future for the Web or Television Shopping?" *Journal of Vacation Marketing* 7(2): 110–124.

Morrison, Alastair. 2002. "Tourism, Hospitality and Leisure Journals." Accessed April 7, 2010. http://omni.cc.purdue.edu/~alltson/journals.htm.

Morrison, Alastair M. 2002. *Hospitality and Travel Marketing*. 3rd ed. Albany: Delmar Cengage Learning.

Morrison, Alastair M., Juline E. Mills, Suttichai Chuvessiriporn, and Joseph A. Ismail. 2002. "Where Are We Now? An Initial Analysis of Web-Based Marketing Issues Affecting Travel and Tourism." In *Information and Communications Technologies in Tourism 2002*, edited by Karl Wober, Andrew Frew, and Martin Hitz, 375–386. New York: Springer Verlag.

Morrison, Alison. 1994. "Marketing Strategic Alliances: The Small Hotel Firm." *International Journal of Contemporary Hospitality Management* 6(3): 25–30.

Morrison, Alison, Michael Rimmington, and Clare Williams. 1999. *Entrepreneurship in the Hospitality, Tourism and Leisure Industries*. Oxford: Butterworth-Heinemann.

Morrison, Margaret. 2002. *Using Qualitative Research in Advertising: Strategies, Techniques, and Applications*. Thousand Oaks, CA: Sage.

Morse, Janice M. 2003. "Principles of Mixed Methods and Multimethod Research Design." In *Handbook of Mixed Methods in Social and Behavioral Research*, edited by Charles Teddlie, and Abbas Tashakkori, 189-208. London: Sage.

Morton, Michael Scott. 1991. *The Corporation of the 1990s—Information Technology and Organizational Transformation*. New York: Oxford University Press.

Mosaic Group. 1998. "The Global Diffusion of the Internet Project: An Initial Inductive Study." Accessed January 16, 2010. www.agsd.com/gdi97/gdi97.html.

Moutinho, Luiz. 1987. "Consumer Behaviour in Tourism." *European Journal of Marketing* 21: 5–44.

Mowery, David C., Joanne E. Oxley, and Brian Silverman. 1998. "Technological Overlap and Interfirm Cooperation: Implications for the Resource-Based View of the Firm." *Research Policy* 27: 507–523.

Mpofu, K.C., D. Milne, and L. Watkins-Mathys. 2013. "ICT Adoption and Development of E-business among SMEs in South Africa." Accessed May 9, 2020. https://api.semanticscholar.org/CorpusID:55965664.

Munro, Barbara H. 2001. *Statistical Methods for Health Care Research*. 4th ed. Philadelphia: Lippincott.

Murphy, William H. 1999. "Hofstede's National Culture as a Guide for Sales Practices across Countries: The Case of a MNC's Sales Practices in Australia and New Zealand." *Australian Journal of Management* 24(1): 37–58.

Nelson, Richard R., and Sidney Winter. 1982. *An Evolutionary Theory of Economic Behaviour and Capabilities*. Cambridge: Harvard University Press, 195–307.

Neuman, W. Lawrence. 2006. *Social Research Methods: Qualitative and Quantitative Approaches*. 6th ed. Boston: Allyn and Bacon.

Nolan, Sidney D. 1976. "Tourists' Use and Evaluation of Travel Information Sources: Summary and Conclusions." *Journal of Travel Research*14(2): 6–8.

Norris, Pipps. 2000. "Digital Divide? Civic Engagement, Information Poverty and the Internet in Democratic Societies." Accessed November 9, 2010.

Northouse, P.G. 2018. *Leadership: Theory and Practice*. Thousand Oaks: Sage.

Norum, Pamela S. 2008. "Student Internet Purchases." *Family and Consumer Sciences Research Journal* 36: 373–388.

NUA. 2003. *Internet Demographics*.

Nysveen, Herbjørn, and Maria Lexhagen. 2002. "Reduction of Perceived Risk through On-line Value-Added Services Offered on Tourism Business Websites." In *Information and Communications Technologies in Tourism 2002*, edited by Karl Wober, Andrew Frew, and Martin Hitz, 488–496. New York: Springer Verlag.

O'Connor, Peter. 2000. *Evaluating Electronic Channels of Distribution in the Hotel Sector*. Barcelona: Springer Verlag.

———. 2001. "The Changing Face of Hotel Electronic Distribution." *Travel and Tourism Analyst* 5: 61–78.

———. 2002. *An Analysis of the Online Pricing Strategies of the International Hotel Chains*. Innsbruck: Springer Verlag.

———. 2008, January. "User-Generated Content and Travel: A Case Study on Tripadvisor.com." In *ENTER*, vol. 2008, 47–58.

O'Connor, Peter, Dimitrios Buhalis, and Andrew Frew. 2001. "The Transformation of Tourism Distribution Channels through Information Technology." In *Tourism Distribution Channels: Practices, Issues and Transformations*, edited by Dimitrios Buhalis, and Eric Laws, 332–350. New York: Continuum.

OECD. 2001. "Organization for Economic Co-operation and Development: Measuring Electronic Commerce." OECD 1997. Accessed December 4, 2010. http://www.oecd.org/dsti/sti/it/ec/prod/e_97-185-htm.

Öörni, Anssi. 2004. "Consumer Objectives and the Amount of Search in Electronic Travel and Tourism Markets." *Journal of Travel and Tourism Marketing* 17(2/3): 3–14.

Opodo. 2003. www.opodo.com.

Opperman, Martin. 1999. "Databased Marketing by Travel Agencies." *Journal of Travel Research* 37(3): 231–237.

Orbitz. 2003. www.orbitz.com.

O'Reilly, Charles A. 1982. "Variations in Decision Makers' Use of Information Sources: The Impact of Quality and Accessibility of Information." *Academy of Management Journal* 25(4): 756–771.

Oz, Effy. 2009. *Managing Information Systems*. 6th ed. South Melbourne, Australia: Cengage Learning.

Page, Stephen. 2005. *Transport and Tourism: Global Perspectives*. 2nd ed. Harlow, England: Pearson, Prentice Hall.

Page, Stephen, and Don Getz. 1997. *The Business of Rural Tourism: International perspectives*. London: Thomson Business Press.

Palmer, A. 1995. "Relationship Marketing: A New Philosophy for Marketing or a New Function?" In *Proceedings of the Seventh Bi-Annual World Marketing Congress*, Vol. 15, 130–134. Melbourne: Monash University.

Palmer, Adrian, and Patrick McCole. 1999. "The Virtual Re-intermediation of Travel Services: A Conceptual Framework and Empirical Investigation." *Journal of Vacation Marketing* 6(1): 33–47.

Pan, Bing, and Daniel Fesenmaier. 2001. "Exploring Caller Dialogue: Analyzing Directed Information Search for Tourism Products." In *Information and Communication Technologies in Tourism 2001: Proceedings of the International Conference in Montreal, Canada, 2001*, edited by Pauline Sheldon, Karl Wober, and Daniel Fesenmaier, 244–250. Wien, Austria: Springer.

———. 2003. "Travel Information Search on the Internet: An Exploratory Study." Doctoral dissertation. Urbana-Champaign: University of Illinois.

Parker, Marilyn. 1983. *Information Economics*. London: Prentice-Hall.

Passmann, Christoph, Wolfgang Pipperger, and Walter Schertler. 1997. "How to Assess WWW- Applications for Tourism Information Systems from the End-User Perspective- Methodological Design and Empirical Evidence." In *Information and Communication Technologies in Tourism: Proceedings of the International Conference in Edinburgh, Scotland*, edited by A. Min Tjoa, 208–219. Wien: Springer.

Patricia, Dodu. 2008. "The Internet, Threat or Tool for Travel Agencies?" Accessed April 9, 2010. http://steconomice.uoradea.ro/anale/volume/2008/v2-economy-andbusiness-administration/017.pdf.

Patton, Michael Quinn. 2002. *Qualitative Evaluation and Research Methods*. 3rd ed. Thousand Oaks, CA: Sage.

Pawar, Badrinarayan S. 2003. "Central Conceptual Issues in Transformational Leadership Research." *Leadership and Organization Development Journal* 24(7): 397–406.

Pawar, Badrinarayan S., and Kenneth Eastman. 1997. "The Nature and Implications of Contextual Influences on Transformational Leadership: A Conceptual Examination." *Academy Management Review* 22: 80–109.

Pearce, D.G., 2008. "A Needs-Functions Model of Tourism Distribution." *Annals of Tourism Research* 35(1): 148–168.

———. 2009. "Beyond Tiers: A Network Approach to Tourism Distribution." *Tourism Analysis* 13: 517–530.

Pennington-Gray, Lori, and Christine Vogt. 2003. "Examining Welcome Center Visitors' Travel and Information Behaviors: Does Location of Centres or Residency Matter?" *Journal of Travel Research* 41: 272–280.

Penrose, Edith T. 1959. *The Theory of the Growth of the Firm*. New York: Wiley.

Perdue, Richard R. 1985. "Segmenting State Information Inquirers by Timing of Destination Decision and Previous Experience." *Journal of Travel Research* 23: 6–11.

———. 1993. "External Information Search in Marine Recreational Fishing." *Leisure Sciences* 15: 169–187.

Perry, Chad A. 1998. "Structured Approach to Presenting Theses." Accessed February 25, 2009. http://www.scu.edu.au/schools/gcm/ar/art/cperry.html.

Persson, Goran. 1978. "Organisation Design Strategies for Business Logistics." *International Journal of Physical Distribution and Materials Management* 8(6): 287–297.

Peteraf, Margaret A. 1993. "The Cornerstones of Competitive Advantage: A Resource-Based View." *Strategic Management Journal* 14: 179–191.

Peterson, Karen Ida. 1994. "Qualitative Research Methods for the Travel and Tourism Industry." In *Travel, Tourism, and Hospitality Research: A Handbook for Managers*, edited by J.R. Brent Ritchie, and Charles R. Goeldner, 487–489. 2nd ed. London: John Wiley and Sons.

Peterson, Suzanne J., Kristin Byron Walumbwa, and Jason Myrowitz. 2009. "CEO Positive Psychological Traits, Transformational Leadership, and Firm Performance in High-technology Start-up and Established Firms." *Journal of Management* 35: 348–368.

Pettigrew, Andrew M. 1987. "Context and Action in the Transformation of Firms." *Journal of Management Studies* 24: 649–670.

PhoCusWright. 2003. *Consumer Trends: Evolution of the Online Traveler*. Sherman: PhoCusWright 2.

Poon, Auliana. 1993. *Tourism, Technology and Competitive Strategies*. Wallingford: CAB International.

———. 2001. "The Future of Travel Agents." *Travel & Tourism Analyst* 3: 57–80.

Poon, Simpson, and Mathew Joseph. 2001. "A Preliminary Study of Product Nature and Electronic Commerce." *Marketing Intelligence & Planning* 19(7): 493–499.

Porter, M.E., 2008. *Competitive Strategy: Techniques for Analyzing Industries and Competitors*.New York: Simon and Schuster.

Porter, Michael. 1980. *Competitive Strategy*. New York: The Free Press.

———. 1985. *Competitive Advantage*. New York: The Free Press.

———. 1990. *The Competitive Advantage of Nations*. London: Macmillan.

———. 1995. *Wettbewerbsstrategien (Competitive Strategy)*. New York: Campus.

———. 2001. "Bewährte Strategien Werden mit dem Internet Noch Wirksamer." *Harvard Business Manager* 5: 64–81.

Porter, Michael, and Victor Millar. 1985. "How Information Gives You Competitive Advantage." *Harvard Business Review* 63(4): 149–160.

Presutti, William. 2003. "Supply Chain Management and E-procurement: Creating Value-added in the Supply Chain." *Industrial Marketing Management* 32: 219–226.

Radford, Gary P. 2005. *On the Philosophy of Communication*. Belmont: Thomson Wadsworth.

Raitz, Karl, and Meftah Dakhil. 1989. "A Note about Information Sources for Preferred Recreational Environments." *Journal of Travel Research* 27: 45–49.

Rao, Subba, Glenn Metts, and Carlo Monge. 2003. "Electronic Commerce Development in Small and Medium Sized Enterprises: A Stage Model and its Implications." *Business Process Management Journal* 9(1): 11–32.

Rappa, Michael. 2002. "Business Models on the Web." Accessed February 7, 2010. http://digitalenterprise.org/models/models.html.

Rayport, Jeffrey F. 1995. "Exploiting the Virtual Value Chain." *Harvard Business Review* 73(6): 75–85.

Rayport, Jeffrey, and Bernard Jaworski. 2002. *Introduction to E-Commerce*. 2nd ed. Boston: Irwin/McGraw-Hill.

Redding, S.Gordon. 1994. "Comparative Management Theory: Jungle, Zoo or Fossil Bed?" *Organization Studies* 15: 323–359.

Reisinger, Yvette, and Lindsay W. Turner. 2002. "Cultural Differences between Asian Tourist Markets and Australian Hosts: Part 2." *Journal of Travel Research* 40: 385–395.

Reve, Torger. 1990. "The Firm as a Nexus of Internal and External Contracts." In *The Firm as a Nexus of Treaties*, edited by Masahiko Aioki, 136–188. London: Sage.

Reynolds, Thomas J., and Jonathan Gutman. 1988. "Laddering Theory, Method, Analysis, and Interpretation." *Journal of Advertising Research* 28: 11–31.

Ricardo, David. 1817. *Principles of Political Economy and Taxation*. London: J. Murray.

Ricketts, Martin. 1994. *The Economies of Business Enterprise*. Hempstead: Harvester Wheatsheaf.

Rideng, Arne, and Berit Grue. 2002. *Foreign Visitor Survey 2002*. Oslo: TØI.

Riley, Roger. 1996. "Revealing Socially Constructed Knowledge through Quasi-structured Interviews and Grounded Theory Analysis." *Journal of Travel and Tourism Marketing* 5(1/2): 21–40.

Riley, Roger W., and Lisa Love. 1999. "The State of Qualitative Tourism Research." *Annals of Tourism Research* 27(1): 164–187.

Rodan, Simon, and Charles Galunic. 2004. "More Than Network Structure: How Knowledge Heterogeneity Influences Managerial Performance and Innovativeness." *Management Journal* 25(6): 541–562.

Roehl, Wesley S., and Daniel Fesenmaier. 1992. "Risk Perceptions and Pleasure Travel: An Exploratory Analysis." *Journal of Travel Research* 30(4): 17–26.

Rogers, E.M. 2010. *Diffusion of Innovations*. Ney York: Simon and Schuster.

Rogers, Everett. 1962. *Diffusion of Innovations*. New York: The Free Press.

———. 1976. "New Product Adoption and Diffusion." *Journal of Consumer Research* 2: 290.

———. 1983. *Diffusion of Innovations*. New York: The Free Press.

———. 1986. *Communication Technology: The New Media in Society*. New York: NY Free Press.

———. 1994. *A History of Communication Study—A Biographical Approach*. New York: The Free Press.

———. 2003. *Diffusion of Innovations*. 5th ed. New York: The Free Press.

Rogers, Everett, and Thomas Valente. 1993. "A History of Information Theory in Communication Research." In *Between Communication and Information—Information and Behavior*, edited by Jorge Reina Schement, and Brent D. Ruben, 35–56. New Brunswick, NJ: Transaction.

Rossman, Gretchen B., and Bruce Wilson. 1985. "Numbers and Words: Combining Quantitative and Qualitative Methods in a Single Large-scale Evaluation Study." *Evaluation Review* 9(5): 627.

Rushton, Alan, and John Oxley. 1993. *Handbook of Logistics and Distribution Management*. London: Kogan Page Limited.

Ryals, Lynette, and Andre Humphries. 2007. "Managing Key Business-to-Business Relationships: What Marketing Can Learn from Supply Chain Management." *Journal of Service Research* 9: 312.

Ryan, Bryce, and Neal Gross. 1943. "The Diffusion of Hybrid Seed Corn in Two Iowa Communities." *Rural Sociology* 8: 15-24.

Saffu, Kojo, and John Walker. 2008. "Strategic Value and Electronic Commerce Adoption among Small and Medium-sized Enterprises in a Transactional Economy." *Journal of Business and Industrial Marketing* 23(6): 395–404.

Sarros, James C., Brain K. Cooper, and Joseph Santora. 2008. "Building a Climate for Innovation through Transformational Leadership and Organizational Culture." *Journal of Leadership & Organizational Studies* 15: 145–158.

Sassi, Sinikka. 2005. "Cultural Differentiation or Social Segregation? Four Approaches to the Digital Divide." *New Media and Society* 7(5): 684–700.

Schary, Philip, and Tage Larsen. 1998. *Managing the Global Supply Chain*. Copenhagen: Copenhagen Business School Press.

Schiefer, Alex. 2002. *eCommerce 2001/2002*. Vienna: Statistik Austria.

Schiffman Leon G., and Leslie Kanuk. 2000. *Consumer Behavior*. Upper Saddle River, NJ: Prentice-Hall.

Schmidt, Jeffrey B., and Richard A. Spreng. 1996. "Proposed Model of External Consumer Information Search." *Academy of Marketing Science Journal* 24(3): 246–270.

Schmitz, Joseph, and Janet Fulk. 1991. "Organizational Colleagues, Media Richness, and Electronic Mail: A Test of the Social Influence Model of Technology Use." *Communication Research* 18(4): 487–523.

Schramm, Wilbur. 1954. "How Communication Works." In *The Process and Effect of Mass Communication*, edited by Wilbur Lang Schramm, 3–26. Urbana: University of Illinois Press.

Schul, Patrick, and John L. Crompton. 1983. "Search Behaviour of International Vacationers: Travel-specific Lifestyle and Sociodemographic Variables." *Journal of Travel Research* 21: 25–31.

Schwartz, Shalom. 1992. "Universals in the Content and Structure of Values: Theoretical Advances and Empirical Tests in 20 Countries." *Advances in Experimental Social Psychology* 25: 1–65.

Schweda, Anika. 2004. "Independent International Traveller Evaluations of Traditional and Interactive Holiday Information Sources along Temporal and Utility Dimensions." In *Information and Communication Technologies in Tourism*, edited by Andrew J. Frew, 151–160. Vienna: Springer Verlag.

Seidman, Irving. 1998. *Interviewing as Qualitative Research—A Guide for Researchers in Educational and Social Sciences*. 2nd ed. New York: Teachers College Press.

Shailer, Gregory. 1994. "Capitalists and Entrepreneurs in Owner-Managed Firms." *International Small Business Journal* 12: 33–41.

Shannon, Claude E., and Warren Weaver. 1963. *The Mathematical Theory of Communication*. Urbana and Chicago: University of Illinois Press.

Sheldon, Pauline. 1997. *Tourism Information Technology*. New York: Cab International.

Sherlock, Philip, and Hazel Bennett. 1998. *The Story of the Jamaican People*. Kingston: Ian Randle Publishers.

Shin, Namchul. 2001. "Strategies for Competitive Advantage in Electronic Commerce." *Journal of Electronic Commerce Research* 2(4): 164–171.

Siebens, H. 2002. "Concepts and Working Instruments for Corporate Governance." *Journal of Business Ethics* 39(1–2): 109–116.

Simon Herbert Alexander. 1957. *Models of Man*. New York: John Wiley and Sons.

Simpson, Christopher. 1994. *Science of Coercion: Communication Research and Psychological Warfare, 1945–1960*. Oxford, England: Oxford University Press.

Sims-Shouten, Wendy, Sarah Riley, and Carla Willig. 2007. "Critical Realism in Discourse Analysis: A Presentation of a Systematic Method of Analysis Using Women's Talk of Motherhood, Childcare and Female Employment as an Example." *Theory and Psychology* 17: 101–124.

Singh, Niti, and R. Venkat Krishnan. 2007. "Transformational Leadership in India: Developing and Validating a New Scale." *International Journal of Cross Cultural Management* 7: 219.

Singh, Surendra P., Ruthie G. Reynolds, and Safdar Muhammad. 2001. "A Gender-based Analysis of Micro and Small Enterprises in Java, Indonesia." *Journal of Small Business Management* 39(2): 174–182.

Slevin, D.P., & J.G. Covin. 1995. "Entrepreneurship as Firm Behavior: A Research Model." *Advances in Entrepreneurship, Firm Emergence, and Growth* 2: 175–224.

Smith, Karen. 2007. "Distribution Channels for Events: Supply and Demand-side Perspectives." *Journal of Vacation Marketing* 13: 321–338.

Smith, L., and P. Mounter. 2008. *Effective Internal Communication*. London: Kogan Page Publishers.

Snepenger, David J. 1987. "Segmenting the Vacation Market by Novelty Seeking Role." *Journal of Travel Research* 26: 8–14.

Snepenger, David J., and Mary Snepenger. 1993. "Market Structure Analysis of Media Selection Practices by Travel Services." In *Communication and Channel Systems in Tourism Marketing*, edited by Sheauhsing Hsieh, and Daniel R. Fesenmaier, 21–36. New Jersey: The Hawthorn Press.

Snepenger, David J., Kelli Meged, Mary Snelling, and Kelly Worrall. 1990. "Information Search Strategies by Destination-naive Tourists." *Journal of Travel Research* 29: 13–16.

Sparks, Colin. 2000. "The Distribution of Online Resources and the Democratic Potential of the Internet." In *Cultural, Economic and Policy Foundations for Media Openness and Diversity in East and West*, edited by Jan van Cuilenburg, and Richard van der Wurff, 229–254. Amsterdam: Het Spinhuis.

Spink, A., and C. Cole. 2001. "Information and Poverty: Information-Seeking Channels Used by African American Low-Income Households." *Library & Information Science Research* 23(1): 45–65.

Standing, Craig, and Thandarayan Vasudavan. 1999. "Internet Marketing Strategies Used by Travel Agencies in Australia." *Journal of Vacation Marketing* 6: 21–32.

Stansfield, Mark, and Kevin Grant. 2003. "An Investigation into Issues Influencing the Use of the Internet and Electronic Commerce among Small-Medium Sized Enterprises." *Journal of Electronic Commerce Research* 4(1): 15–33.

Statistical Institute of Jamaica. Accessed November 14, 2010. http://statinja.gov.jm/population.aspx.

Stone, Carl. 1973. *Class, Race, and Political Behaviour in Urban Jamaica. Institute of Social and Economic Research*. Mona: University of the West Indies.

———. 1985. *Class, State, and Democracy in Jamaica*. Kingston: Blackett Publishers.

Stonehouse, George, and Brian Snowdon. 2007. "Competitive Advantage Revisited: Michael Porter on Strategy and Competitiveness." *Journal of Management Inquiry* 16(3): 256–273.

Straub, Evan. 2009. "Understanding Technology Adoption: Theory and Future Directions for Informal Learning." *Review of Educational Research* 79: 625–649.

Strilka, Bernhard. 2002. *Das Internet als Marketinginstrument im Tourismus - unter besonderer Berücksichtigung der österreichischen Klein- und Mittelbetrieb-sstruktur*. Vienna: University of Business and Economics Administration.

Stump, Rodney L., Wen Gong, and Zhan Li. 2008. "Exploring the Digital Divide in Mobile-phone Adoption Levels across Countries: Do Population Socioeconomic Traits Operate in the Same Manner as Their Individual-level Demographic Counterparts?" *Journal of Macromarketing* 28: 397–412.

Subramaniam, Nava, and Neal Ashkanasy. 2001. "The Effect of Organisational Culture Perceptions on the Relationship between Budgetary Participation and Managerial Job-Related Outcomes." *Australian Journal of Management* 26(1): 35–54.

Swanson, E. Burton. 1987. "Information Channel Disposition and Use." *Decision Sciences* 18: 131–145.

Tabachnick, Barbara G., and Linda Fidell. 2001. *Using Multivariate Statistics*. 4th ed. Boston: Allyn and Bacon.

Tan, S.J. 1999. "Strategies for Reducing Consumers' Risk Aversion in Internet Shopping." *Journal of Consumer Marketing*.

Tan, Wen Dee, and David R. Arnott. 1999. "Managerial Information Acquisition and the World Wide Web: A Theoretical Background." ISDSS '99.

Tang, Lijun, and Peidong Yang. 2011. "Symbolic Power and the Internet: The Power of a 'Horse'." *Media Culture Society* 33: 675–691.

Tangpong, Chanchai, and Nongkran Lertpittaypoom. 2009. "The Emergence of Business-to-Customer E-commerce: New Niche Formation, Creative Destruction and Contingency Perspectives." *Journal of Leadership and Organizational Studies* 16(2): 131–140.

Tashakkori, Abbas, and Charles Teddlie. 2003. "The Past and Future of Mixed Methods Research: From Data Triangulation to Mixed Model Designs." In *Handbook of Mixed Methods in Social and Behavioral Research*, edited by Charles Teddlie, and Abbas Tashakkori, 671–702. London: Sage.

Tatum, B. Charles, Richard Eberlin, Carin Kottraba, and Travis Bradberry. 2003. "Leadership, Decision-Making and Organizational Justice." *Management Decision* 41(10): 1006–1016.

Taylor, David. 1999. *Global Cases in Logistics and Supply Chain Management*. London: International Thomson Business Press.

Tchaïcha, Jane D., and Mark M. Davis. 2005. "The Impact of Culture on Technology and Business: An Interdisciplinary, Experiential Course Paradigm." *Journal of Management Education* 29(5): 738–757.

Tebbutt, Jane. 2006. "E-Business and Small Firms in London." In *Impact of E-Commerce on Consumers and Small Firms*, edited by Salvatore Zappala and Colin Gray, 63–80. Aldershot: Ashgate.

Teece, David J. 1982. "Toward an Economic Theory of the Multiproduct Firm." *Journal of Economic Behaviour and Organization* 3: 39–63.

———. 2010. "Business Models, Business Strategy and Innovation." *Long Range Planning* 43(2–3): 172–194.

Teo, Thompson, Sijie Lin, and Kee-hung Lai. 2009. "Adopters and Non-adopters of E-procurement in Singapore: An Empirical Study." *Omega* 37(5): 972–987.

Teo, T.S. and Y. Pian. 2003. "A Contingency Perspective on Internet Adoption and Competitive Advantage." *European Journal of Information Systems* 12(2): 78–92.

Terpstra, Vern, and Kenneth David. 1991. *The Cultural Environment of International Business*. Cincinnati, OH: South Western Publishing.

Tetzeli, Rick. 1994. "The Internet and Your Business." *Fortune* 129(5): 86–96.

Theobald, Axel, and Marcus Dreyer. 2001. *Online-Marktforschung*. Wiesbaden: Gabler Verlag.

Thomas, Robyn, Cynthia Hardy, and Leisa Sargent. 2011. "Managing Change: Negotiating Meaning and Power-Resistance Relations." *Organization Science* 22(1): 22–41.

Thomas, Robyn, and David Dunkerley. 1999. "Careering Downwards?: Middle Management in the Downsized Organization." *British Journal of Management* 10(2): 157–169.

Thompson, John L. 1990. *Strategic Management*. London: Chapman & Hall.

Thong, James Y.L., and C. S. Yap. 1995. "CEO Characteristics, Organisational Characteristics and Information Technology Adoption in Small Business." *Omega-international Journal of Management Science* 23(4): 429–442.

Tierney, Patrick T. 1993. "The Influence of State Travel Information Centers on Tourist Length of Stay and Expenditure." *Journal of Travel Research* 31(3): 28–32.

Timmers, Paul. 1998. "Business Models for Electronic Markets." *Electronic Markets* 8: 3–8.

Tiscover. 2003. "Tiscover Company Profile." www.aboutiscover.com.

Tjostheim, Ingvar. 1997. "The New Network in the Norwegian Tourism Industry- Marketing and Organisational Theories as Perspectives for a Study." In *Information and Communication Technologies in Tourism: Proceedings of the International Conference in Edinburgh, Scotland, 1997*, edited by A. Min Tjoa, 110–118. Wien: Springer.

———. 2002. "The Internet in Competition with the Traditional Information Sources—A Study of Vacation Planning for Package vs. Non-package Travel." In *Information and Communications Technologies in Tourism 2002*, edited by Karl Wober, Andrew J. Frew, and Martin Hitz, 166–176. New York: Springer Verlag.

Tomasetto, Carlo, and Felice Carugati. 2006. "Social Influence and Diffusion of Innovations in Education." In *Impact of E-Commerce on Consumers and Small Firms*, edited by Salvatore Zappala, and Colin Gray, 21–34. Aldershot: Ashgate.

Toomela, Aaro. 2003. "How Should Culture Be Studied?" *Culture & Psychology* 9(1): 35–45.

Trenholm, Sarah. 1991. *Human Communication Theory*. 2nd ed. Englewood Cliffs, NJ: Prentice Hall.

Trevino, Linda K., Richard L. Daft, and Robert Lengel. 1990. "Understanding Managers' Media Choices: A Symbolic Interactionist Perspective." In *Organizations and Communication Technology*, edited by Janet Fulk, and Charles W. Steinfield, 71–94. Newbury Park, CA: Sage.

Trochim, William M. K. 2006. "Research Methods Knowledge Base." Accessed November 17, 2009. www.socialresearchmethods.net/kb/positivism.php.

Trompenaars, Fons. 1993. *Riding the Waves of Culture: Understanding Cultural Diversity in Business*. London: Economist Books.

Tsai, Henry, and Zheng Gu. 2007. "Institutional Ownership and Firm Performance: Empirical Evidence from U.S. Based Publicly Traded Restaurant Firms." *Journal of Hospitality and Tourism Research* 31(1): 19–38.

Tsao, Chiung-Wen. 2009. "Founding-family Ownership and Firm Performance: The Role of High-performance Work Systems." *Family Business Review* 22(4): 319–332.

Tylor, Edward. 1897. *Primitive Culture*. New York: J.P. Putnam's Sons.

Ulijn, Jan, Dan O'Hair, Mathieu Weggeman, Gerald Ledlow, and Thomas Hall. 2000. "Innovation, Corporate Strategy, and Cultural Context: What Is the Mission for International Business Communication?" *Journal of Business Communication* 37(3): 293–317.

Um, Seoho, and John L. Crompton. 1990. "Attitude Determinants in Tourism Destination Choice." *Annals of Tourism Research* 17: 432–448.

Uncles, M.D., Dowling, G.R. and Hammond, K., 2003. "Customer Loyalty and Customer Loyalty Programs." *Journal of Consumer Marketing* 20: 294–316.

U.S. Department of Commerce. 2007. "Quarterly Retail E-commerce Sales: 1st Quarter 2007 U.S. Census Bureau News." Accessed March 17, 2009. www.commerce.gov.

Uysal, Muzaffer, Cary D. McDonald, and Laurel Reid. 1990. "Sources of Information Used by International Visitors to U.S. Parks and Natural Areas." *Journal of Park and Recreation Administration* 8(1): 51–59.

Van Raaij, W. Fred. 1986. "Consumer Research on Tourism: Mental and Behavioural Constructs." *Annals of Tourism Research* 13: 1–9.

Van Raaij, W. Fred, and Dick A. Francken. 1984. "Vacation Decisions, Activities, and Satisfactions." *Annals of Tourism Research* 11(1): 101–112.

Vaughan, Roger. 2008. "Conceptual Framework." Power Point Presentation, Bournemouth University. Accessed May 14, 2010. http://www.powershow.com/view/a57f7-NDM1Y/Conceptual_Framework_Professor_Roger_Vaughan_May_29th_2008_flash_ppt_presentation.

Veal, Anthony James. 1997. *Research Methods for Leisure and Tourism*. 2nd ed. London: Pitman Publishing.

Venkatesh, Viswanath, and Fred Davis. 2000. "A Theoretical Extension of the Technology Acceptance Model: Four Longitudinal Field Studies." *Management Science* 46(2): 186–204.

Victorino, Liana, Rohit Verma, Gerhard Plaschka, and Chekitan Dev. 2006. "Service Innovation and Customer Choices in the Hospitality Industry." *Managing Service Quality* 15(6): 555–576.

Villalonga, B., and R. Amit. 2006. How Do Family Ownership, Control and Management Affect Firm Value? *Journal of Financial Economics* 80(2): 385–417.

Virolainen, Veli. 1998. "A Survey of Procurement Management Development in Industrial Companies." *International Journal of Productions Economics* 56: 677–688.

Vishwanath, Arun, and Gerald Goldhaber. 2003. "An Examination of the Factors Contributing to Adoption Decisions among Late-Diffused Technology Products." *New Media and Society* 5(4): 547–572.

Vogt, Christine A., and Daniel R. Fesenmaier. 1998. " Expanding the Functional Information Search Model." *Annals of Tourism Research* 25(3): 551–578.

Vogt, Christine A., Susan I. Stewart, and Daniel R. Fesenmaier. 1998. "Communication Strategies to Reach First-time Visitors." *Journal of Travel and Tourism Marketing* 7(2): 69–89.

Vrana, Vasiliki, and Costas Zafiropoulos. 2006. "Tourism Agents' Attitudes on Internet Adoption: An Analysis from Greece." *International Journal of Contemporary Hospitality Management* 18(7): 601–608.

Wacquant, Loic J.D. 1996. "The Rise of Advanced Marginality. Notes on Its Nature and Implications." *Acta Sociologica* 39(2): 121–139.

Waldman, David A., Bernard M. Bass, and Francis Yammarino. 1990. "Adding to Contingent Reward Behavior." *Group and Organisation Studies* 15: 382–394.

Walle, Alf H. 1997. "Quantitative Versus Qualitative Tourism Research." *Annals of Tourism Research* 24(3): 524–536.

Walters, Bruce A., Mark Kroll, and Peter Wright. 2008. "CEO Ownership and Effective Boards: Impacts on Firm Outcomes." *Strategic Organization* 6(3): 259–283.

Warden, Stuart, and S. Tunzelana. 2004. "E-Commerce: A Critical Review of SMME Organizational Barriers in Tourism." Paper presented at the Proceedings of 6th WWW Applications Conference 2004, Rand Afrikaans University, Johannesburg, South Africa.

Wasilczuk, Julita. 2000. "Advantageous Competence of Owner/ Managers to Grow the Firm in Poland: Empirical Evidence." *Journal of Small Business Management* 38(2): 88–94.

Weis, Hans C., and Peter Steinmetz. 2000. *Martkforschung*. Ludwigshafen: Kiehl Verlag.

Weiss, Michael J. 2003. "To Be About to Be." *American Demographics* 25(7): 29–36.

Welch, Emma. 2003. "The Relationship between Ownership Structure and Performance in Listed Australian Companies." *Australian Journal of Management* 28(3): 287–306.

Weller, Susan C., and Kimball Rommey. 1988. *Systematic Data Collection*. Newbury Park, California: Sage.

Wellner, Alison S. 2000. "Generational Divide." *American Demographics* 22(10): 52–58.

Wernerfelt, Birger. 1984. "A Resource-Based View of the Firm." *Strategic Management Journal* 5: 171–180.

Werthner, Hannes. 1995a. *Design Principles of Tourism Information Systems*. Wien, Oldenburg.

———. ed. 1995b. *Elektronisches Marketing im Tourismus*. Wien: Oldenburg.

———. 1996. *Design Principles of Tourism Information Systems*. Innsbruck: Springer Verlag.

Werthner, Hannes, and Stefan Klein. 1999a. *Information Technology and Tourism: A Challenging Relationship*. Vienna: Springer Verlag.

———. 1999b. "ICT and the Changing Landscape of Global Tourism Distribution." *Electronic Markets* 9(4): 256–262.

Westwood, Robert, and David Low. 2003. "The Multicultural Muse Culture, Creativity and Innovation." *International Journal of Cross Cultural Management* 3(2): 235–259.

Whyte, Millicent. 1983. *A Short History of Education in Jamaica*. London: Hodder and Stoughton.

Wicks, Bruce E., and Michael Schuett. 1993. "Using Travel Brochures to Target Frequent Travellers and "Big-spenders"." In *Communication and Channel Systems in Tourism Marketing*, edited by Sheauhsing Hsieh, and Daniel R. Fesenmaier, 77–89. The Hawthorn Press.

Wiertz, Caroline. 2001. "The Internet Adoption Decision in Small and Medium Enterprises (SMEs)." Accessed February 12, 2010. http://www.bonissen.de/ib/papers/Small%20Biz.pdf.

Willcocks, Leslie. 2000. *Moving to E-business: The Ultimate Guide to Effective E-Business*. London: Random House Business Book.

Williams, Densil, and Andrew Spencer. 2010 "Advertising and Tourist Arrivals: Evidence from Jamaica." *Tourism and Hospitality Research* 10(4): 359–366.

Williamson, Dermot. 2002. "Forward from a Critique of Hofstede's Model of National Culture." *Human Relations* 55(11): 1373–1395.

Williamson, Oliver E. 1979. "Transaction Cost-Economies: The Governance of Contractual Relations." *Journal of Law and Economics* 22: 232–261.

———. 1985. *The Economic Institutions of Capitalism*. New York: Free Press.

———. 1990. "The Firm as a Nexus of Treaties: An Introduction." In *The Firm as a Nexus of Treaties*, edited by Masahiko Aioki, 136–188. London: Sage.

Williamson, Oliver E., and Scott E. Masten. 1999. *The Economics of Transaction Costs*. Cheltenham: Edward Elgar Publishing.

Willis, Suzanne, and Bruce Tranter. 2006. "Beyond the 'Digital Divide': Internet Diffusion and Inequality in Australia." *Journal of Sociology* 42(1): 43–59.

Wilson, Kenneth R., Jennifer S. Wallin, and Christa Reiser. 2003. "Social Stratification and the Digital Divide." *Social Science Computer Review* 21: 133–143.

Winston, Brian. 1986. *Misunderstanding Media*. London: Routledge and Kegan Paul.

Wirtz, Bernd W. 2001. *Electronic Business*. Wiesbaden: Gabler Verlag.

Wöber, Karl. 2003. "Information Supply in Tourism Management by Marketing Decision Support Systems." *Tourism Management* 24: 241–255.

Wöber, Karl, and Andrew Frew. 2002. *Information and Communication Technologies in Tourism 2002*. Innsbruck: Springer Verlag.

Wöber, Karl, and Ulrike Gretzel. 2000. "Tourism Managers' Adoption of Marketing Decision Support Systems." *Journal of Travel Research* 39: 172–181.

Woiceshyn, Jaana. 2000. "Technology Adoption: Organizational Learning in Oil Firms." *Organization Studies* 21(6): 1095–1118.

Wolfe, Kara, Cathy H. C. Hsu, and Soo Kang. 2004. "Buyer Characteristics among Users of Various Travel Intermediaries." *Journal of Travel and Tourism Marketing* 17(2/3): 51–62.

Woodside, Arch G., and Ilkka Ronkainen. 1980. "Vacation Travel Planning Segments: Self Planning Vs. Users of Motor Club and Travel Agents." *Annals of Tourism Research* 7: 385–393.

Woodward, Tracy. 2000. "Using Brand Awareness and Brand Image in Tourism Channels of Distribution." *Journal of Vacation Marketing* 6: 119–130.

World Travel and Tourism Council. 2019. *Travel & Tourism-Economic Impact 2019-World*.

WTO. 2000. *World Tourism Organisation's Tourism 2020*. Madrid: World Tourism Organisation.

WTTC. 2002. *The Tourism Satellite Account Method of the World Travel and Tourism Council*. London: WTTC.

Wynne, Clive, Pierre Berthon, Leyland Pitt, Michael Ewing, and Julie Napoli. 2001. "The Impact of the Internet on the Distribution Value Chain: The Case of the South African Tourism Industry." *International Marketing Review* 18(4): 420–431.

Xue, Susan. 2005. "Internet Policy and Diffusion in China, Malaysia and Singapore." *Journal of Information Science* 31(3): 238–250.

Yin, Robert K. 1994. *Case Study Research: Design and Methods*. 2nd ed. Newbury Park, CA: Sage.

Ylänne-Mcewen, Virpi. 2004. "Shifting Alignment and Negotiating Sociality in Travel Agency Discourse." *Discourse Studies* 6: 517–536.

Youn, Seounmi H. 2000. "The Dimensional Structure of Consumer Buying Impulsivity: Measurement and Validation." PhD Diss., University of Minnesota.

Yu, Liangzhi. 2006. "Understanding Information Inequality: Making Sense of the Literature of the Information and Digital Divides." *Journal of Librarianship and Information Science* 38(4): 229–252.

Yu, A. K. L., and Rob Law. 2000. *The Application of eCommerce to Enhance the Competitive Advantage of Hotels in Hong Kong*. Barcelona: Springer Verlag.

Yun, Jin-Hyo, Sangmoon Park, and Mohan Avvari. 2011. "Development and Social Diffusion of Technological Innovation: Cases Based on Mobile Telecommunications in National Emergency Management." *Science Technology Society* 16: 215–234.

Zaichkowsky, Judith L. 1985. "Measuring the Involvement Construct." *Journal of Consumer Research* 12(3): 341–352.

Zappala, Salvatore, and Colin Gray. 2006. *Impact of e-Commerce on Consumers and Small Firms*. Aldershot: Ashgate.

Zappala, Salvatore, and Guido Sarchielli. 2006. "Climate for Innovation, Attitude to Internet and ICT Adoption in Small Firms." In *Impact of e-Commerce on Consumers and Small Firms*, edited by Salvatore Zappala, and Colin Gray, 35–50. Aldershot: Ashgate.

Zhou, Yuqiong. 2008. "Voluntary Adopters Versus Forced Adopters: Integrating the Diffusion of Innovation Theory and the Technology Acceptance Model to Study Intra-organization Adoption." *New Media Society* 10: 475.

Zhu, F., and X. Zhang. 2010. "Impact of Online Consumer Reviews on Sales: The Moderating Role of Product and Consumer Characteristics." *Journal of Marketing*, 74(2): 133–148.

Zipf, George K. 1949. *Human Behavior and the Principle of Least Effort*. Boston, MA: Addison-Wesley.

Zmud, Robert W. 1987. "An Empirical Investigation of the Dimensionality of the Concept of Information." *Decision Sciences* 9: 187–195.

Index

A. Spencer, *Technology Adoption in the Caribbean Tourism Industry*, https://doi.org/10.1007/978-3-030-61584-0

The manufacturer's authorised representative in the EU is Springer Nature Customer Service Centre GmbH, Europaplatz 3, 69115 Heidelberg, Germany. If you have any concerns regarding our products, please contact ProductSafety@springernature.com

Printed and bound by CPI Group (UK) Ltd, Croydon, CR0 4YY
30/09/2026
02225689-0001